Army Techniques Publication
No. 4-02.2

Headquarters
Department of the Army
Washington, D.C., 12 July 2019

Medical Evacuation

Contents

		Page
	PREFACE	v
	INTRODUCTION	vii
Chapter 1	**THE ARMY HEALTH SYSTEM AND MEDICAL EVACUATION**	**1-1**
	Section I — Army Health System	1-1
	Section II — Medical Evacuation	1-1
	Primary Tasks	1-3
	Attributes	1-3
	Medical Evacuation Tools	1-4
	Section III — Medical Evacuation Versus Casualty Evacuation	1-6
	Casualty Evacuation	1-6
	Mortuary Affairs	1-7
Chapter 2	**ARMY MEDICAL EVACUATION**	**2-1**
	Section I — Medical Evacuation Support	2-1
	Section II — Medical Evacuation Requests	2-2
	Ground Evacuation Request	2-4
	Joint and Multinational Interconnectivity	2-8
	Section III — Medical Evacuation Units, Elements, and Platform Considerations	2-9
	Medical Evacuation Support Protocol	2-9
	Medical Evacuation Platform Factors	2-9
	Section IV — Medical Evacuation at Unit Level	2-10
	Medical Evacuation Mission Considerations	2-10
	Medical Evacuation Support from Role 1 and Role 2 Medical Treatment Facilities	2-12
	Section V — Exchange of Property	2-14
	Section VI — Key Activities During Specific Operations	2-14
	Key Activities During Operations to Shape	2-14
	Key Activities During Operations to Prevent	2-15
	Key Activities During Large-Scale Combat Operations	2-15
	Key Activities During Operations to Consolidate Gains	2-16

DISTRIBUTION RESTRICTION: Approved for public release; distribution is unlimited.

*This publication supersedes ATP 4-02.2, dated 12 August 2014.

Contents

Section VII — Medical Evacuation Support for Operations Conducting Offense, Defense, and Stability Tasks 2-16
Medical Evacuation Support for the Offense 2-16
Medical Evacuation Support for the Defense 2-19
Medical Evacuation Support for Operations Dominated by Stability Tasks 2-20

Section VIII — Medical Evacuation Support for Army Special Operations Forces 2-20

Section IX — Medical Evacuation in Urban Operations 2-22
Urban Terrain and Environment 2-22
Aeromedical Evacuation for Urban Operations 2-24

Section X — Army Support to Civil Authorities 2-24
Defense Support of Civil Authorities 2-24
Medical Evacuation Support for Foreign Humanitarian Assistance 2-26

Section XI — Other Types of Medical Evacuation Support Missions 2-27
Evacuation of Military Working Dogs 2-27
Personnel Recovery Operations 2-27
Shore-to-Ship Evacuation Operations 2-27
Medical Evacuation of Detainees 2-28

Section XII — Medical Evacuation in Specific Environments 2-28
Mountain Operations 2-28
Jungle Operations 2-30
Desert Operations 2-31
Operations in Cold Regions 2-32
Chemical, Biological, Radiological, and Nuclear Contaminated Environments 2-32
Battlefield Obscuration 2-34

Chapter 3 MEDICAL EVACUATION RESOURCES 3-1

Section I — Maneuver Battalion Medical Platoon Ambulance Squad 3-1
Medical Treatment Platoon Ambulance Squads 3-1
Ambulance or Evacuation Squad 3-1
Ambulance Team 3-1

Section II — Evacuation Platoon—Medical Company (Brigade Support Battalion) and Ambulance Platoon—Medical Company (Area Support) 3-2
Medical Companies 3-2
Medical Company (Brigade Support) 3-2
Medical Company (Area Support) 3-3
Organizations 3-3
Ambulance/Evacuation Platoon 3-3
Ambulance Squad 3-3

Section III — Medical Company (Ground Ambulance) 3-4
Mission 3-5
Assignment 3-5
Employment 3-5
Capabilities 3-5
Dependency 3-5
Organization and Functions 3-5

Section IV — Medical Company (Air Ambulance) 15 HH-60 3-6
Mission 3-6
Assignment 3-6
Employment 3-6
Capabilities 3-6
Dependency 3-7
Organization and Functions 3-7

	Section V — Medical Company (Air Ambulance) UH-72 Light Utility Helicopter	3-7
	Mission	3-8
	Assignment	3-8
	Employment	3-8
	Capabilities	3-8
	Dependency	3-9
	Organization and Functions	3-9
Chapter 4	**THEATER EVACUATION POLICY**	**4-1**
	Establishing the Theater Evacuation Policy	4-1
	Factors Determining the Theater Evacuation Policy	4-2
	Impact of the Evacuation Policy on Army Health System Requirements	4-3
Chapter 5	**OPERATIONAL AND TACTICAL EVACUATION PLANNING**	**5-1**
	Section I — Theater Medical Evacuation Planning Responsibilities	**5-1**
	Joint Planning	5-1
	Medical Command and Control Organizations	5-1
	Section II — Planning Process	**5-3**
	Evacuation Plans and Orders	5-3
	Characteristics	5-3
	Plans and Orders	5-3
	Medical Evacuation Planning	5-4
Chapter 6	**MEDICAL REGULATING**	**6-1**
	Purpose of Medical Regulating	6-1
	Medical Regulating Terminology	6-1
	Medical Regulating for the Echelons Above Brigade	6-4
	Medical Regulating Within the Operational Area	6-5
	Medical Regulating from the Operational Area to Echelons Above Brigade	6-5
	Medical Regulating Within Echelons Above Brigade	6-6
	Intertheater Medical Regulating	6-6
	En Route Patient Staging System	6-7
	Limitation of the United States Air Force Theater Aeromedical Evacuation System	6-8
	Originating Medical Facility's Responsibilities	6-9
	Medical Regulating for Army Special Operations Forces	6-9
Appendix A	**GENEVA CONVENTIONS AND THE LAW OF WAR**	**A-1**
Appendix B	**EXAMPLE OF THE MEDICAL EVACUATION PLAN AND OPERATIONS ORDER**	**B-1**
Appendix C	**EXAMPLE OF THE MEDICAL EVACUATION REQUEST**	**C-1**
Appendix D	**EXAMPLE OF THE MEDICAL EVACUATION ACTIVITIES DURING OPERATIONS**	**D-1**
	GLOSSARY	**Glossary-1**
	REFERENCES	**References-1**
	INDEX	**Index-1**

Figures

Figure 2-1. Flow of communication for medical evacuation request 2-4
Figure 2-2. Ground evacuation request in a maneuver unit 2-5

Figure 2-3. Ground evacuation request in the echelon above brigade 2-6
Figure 2-4. Aeromedical evacuation request 2-7
Figure 2-5. Air ambulance zones of evacuation 2-8
Figure 2-6. Air and ground ambulance evacuation in a joint environment 2-9
Figure 3-1. Medical company, ground ambulance 3-4
Figure 3-2. Medical company (air ambulance) 15 HH-60 3-6
Figure 3-3. Medical company (air ambulance) (UH-72 light utility helicopter) 3-8
Figure 5-1. Mission command organizations for coordination and orders flow 5-2
Figure B-1. Example of a medical evacuation plan B-1
Figure B-2. Example of a medical evacuation tab to an AHS appendix B-6
Figure D-1. Medical evacuation support during operations to shape D-2
Figure D-2. Medical evacuation during operations to prevent D-3
Figure D-3. Medical evacuation during large-scale combat operations D-4
Figure D-4. Medical evacuation during consolidation of gains D-5

Tables

Introductory Table-1. Army terms viii
Table 1-1. Primary tasks and purposes of medical evacuation 1-3
Table 2-1. Categories of evacuation precedence 2-2
Table 5-1. Medical evacuation planning considerations 5-5
Table C-1. 9-Line Medical Evacuation Request C-1

Preface

Army Techniques Publication 4-02.2 provides doctrine and techniques for conducting medical evacuation and medical regulating operations. Medical evacuation encompasses both the evacuation of Soldiers from the point of injury or wounding to a medical treatment facility staffed and equipped to provide essential care in theater and further evacuation from the theater to provide definitive, rehabilitative, and convalescent care in the continental United States and the movement of patients between medical treatment facilities or to staging facilities. Medical evacuation entails the movement of patients on dedicated ground and air ambulances, medically staffed and equipped to provide en route medical care; supports the military health system; and links the continuum of care. In addition, it discusses the difference between medical evacuation and casualty evacuation as well as coordination requirements for and the use of nonmedical transportation assets to accomplish the casualty evacuation mission.

The principal audience for this publication is all commanders and their staffs, command surgeons, Army health systems planners, Army Medical Department personnel and units involved in medical evacuation operations.

Commanders, staffs, and subordinates ensure that their decisions and actions comply with applicable United States, international, and in some cases host-nation laws and regulations. Commanders at all levels ensure that their Soldiers operate in accordance with the law of war and the rules of engagement. (See Field Manual 27-10 and Department of Defense Law of War Manual).

Department of Defense Directive 5100.01 directs the United States Army to develop concepts, tactics, techniques, and procedures and organize, train, equip, and provide forces with expeditionary and campaign qualities to provide intratheater aeromedical evacuation.

Army Techniques Publication 4-02.2 implements or is in consonance with the following North Atlantic Treaty Organization Standardization Agreements and American, British, Canadian, Australian and New Zealand (Armies).

TITLE	NATO STANAGs	ABCA Standards
Minimum Labelling Requirements for Medical Materiel		436
Stretchers, Bearing Brackets and Attachment Supports—Allied Medical Publication-2.1	2040	
Forward Aeromedical Evacuation—Allied Aeromedical Publication-1.5	2087	
ABCA Patient Medical Evacuation Request		2079
Coalition Casualty Regulating Tool (CRRT)		2080
Medical and Dental Supply Procedures—Allied Medical Publication-1.12	2128	
Documentation Relative to Initial Medical Treatment and Evacuation—Allied Medical Publication-8.1	2132	
Road Movements and Movement Control—Allied Movement Publication-1(A)	2454	
Orders for the Camouflage of Protective Medical Emblems on Land in Tactical Operations—Allied Tactical Publication-79	2931	
Aeromedical Evacuation—Allied Aeromedical Publication-1.1	3204	

Preface

Army Techniques Publication 4-02.2 uses joint terms where applicable. Selected joint and Army terms and definitions appear in both the glossary and the text. Terms for which Army Techniques Publication 4-02.2 is the proponent publication (the authority) are marked with an asterisk (*) in the glossary. Definitions for which this publication is the proponent publication are boldfaced in the text. For other definitions shown in the text, the term is italicized and the number of the proponent publication follows the definition

This publication applies to the Active Army, Army National Guard/Army National Guard of the United States, and United States Army Reserve unless otherwise stated.

The proponent and preparing agency of Army Techniques Publication 4-02.2 is the United States Army Medical Department Center and School, Health Readiness Center of Excellence. Send comments and recommendations on Department of the Army Form 2028 (Recommended Changes to Publications and Blank Forms) to **Commander, United States Army Medical Department Center and School United States Army Health Readiness Center of Excellence, Attention: MCCS-FD (Army Techniques Publication 4-02.2), Building 4011, Suite D, 2377 Greeley Road, JBSA Fort Sam Houston, TX 78234-7731**; by e-mail to usarmy.jbsa.medcom-ameddcs.mbx.ameddcs-medical-doctrine@mail.mil; or submit an electronic Department of the Army Form 2028. All recommended changes should be keyed to a specific page, paragraph, and line number. A rational for each proposed change is required to aid in the evaluation and adjudication of each comment.

Unless this publication states otherwise, masculine nouns and pronouns do not refer exclusively to men.

Introduction

Army Techniques Publication 4-02.2 is consistent with Field Manual 4-02, and Joint Publication 4-02 while adopting concepts as necessary and updated terminology. The Army Health System is a complex system of interrelated and interdependent systems which provides a continuum of medical treatment from the point of injury or wounding through successive roles of medical care to definitive, rehabilitative, and convalescent care in the continental United States, as required. Medical evacuation is the system which provides the vital linkage between the roles of care necessary to sustain the patient during transport. This is accomplished by providing en route medical care and emergency medical intervention, if required, to enhance the individual's prognosis and reduce long-term disability.

Summary of changes include—
- Reorganizing the order of the publication.
- Designating this publication as the proponent for seven terms for which FM 4-02 was previously the proponent. (See Introductory Table-1.)
- Aligning this publication with FM 3-0 and ADP 4-0.

Army Techniques Publication 4-02.2 contains six chapters and four appendices as follows:

Chapter 1 provides an overview of Army Health System and how medical evacuation relates to the principles of the Army Health System. It discusses the purpose, primary tasks, and attributes of the Army medical evacuation system. It also defines the differences between medical evacuation and casualty evacuation.

Chapter 2 discusses the employment of medical evacuation resources and the coordination and synchronization required to effectively execute medical evacuation operations. This includes the medical evacuation request process, consideration for evacuation missions, support planning considerations, and evacuation in specific environments.

Chapter 3 describes the mission, function and capabilities of medical evacuation units and elements as specified in the unit's table of organization and equipment. It also discusses the mission command headquarters to which they are assigned.

Chapter 4 discusses the factors that establish the evacuation policy and the impact of the evacuation policy on Army Health System support.

Chapter 5 provides insight and considerations into developing the operational and tactical medical evacuation plan that supports the combatant commander's mission.

Chapter 6 describes the medical regulating system designed to ensure the efficient and safe movement of regulated patients to the appropriate military treatment facility by the most effective means. It also discusses the multi-Service responsibility and assets used to conduct this mission.

Appendix A provides a summary of the Geneva Conventions and The Law of War.

Appendix B provides an example of a medical evacuation plan as part of an operations order.

Appendix C provides an example of the 9-line medical evacuation request format.

Appendix D provides examples of medical evacuation during operations to shape, prevent, large-scale combat operations, and to consolidate gains.

Based on doctrinal changes, terms for which Army Techniques Publication 4-02.2 is the proponent have been added for the purposes of this publication. The glossary contains acronyms and defined terms. (See introductory Table-1 on page viii for specific term changes.)

Introductory Table-1. Army terms

Term	Remarks
casualty collection point	ATP 4-02.2 is now the proponent.
en route care	ATP 4-02.2 is now the proponent
lines of patient drift	ATP 4-02.2 is now the proponent
medical evacuation	ATP 4-02.2 is now the proponent
nontransportable patient	ATP 4-02.2 is now the proponent
patient movement	ATP 4-02.2 is now the proponent
theater evacuation policy	ATP 4-02.2 is now the proponent
Legend: ATP Army Techniques Publication	

Chapter 1

The Army Health System and Medical Evacuation

The Army Health System (AHS) is the Army component of the Department of Defense (DOD) Military Health System. Its capabilities are focused on delivering medical care across the range of military operations—from the point of injury (POI) or wounding, through the roles of care within the joint operations area, to the continental United States (CONUS)-support base. Medical evacuation is the key factor in ensuring the continuity of care provided to our Soldiers by providing en route medical care during evacuation, facilitating the transfer of patients between medical treatment facilities (MTFs) to receive the appropriate specialty care, and ensuring that scarce medical resources (personnel, equipment, and supplies [to include blood]) can be rapidly transported to areas of critical need on the battlefield.

SECTION I — ARMY HEALTH SYSTEM

1-1. The AHS supports unified land operations through the provision of health service support (HSS) (casualty care, medical evacuation (MEDEVAC), and medical logistics) as part of the sustainment warfighting function and force health protection (FHP) under the protection warfighting function. The AHS is focused on promoting wellness, preventing casualties due to disease and nonbattle injuries, and providing timely and effective casualty management and care. The provision of AHS support is governed by well-established and time-tested principles and rules which ensure the care provided to our Soldiers is timely and effective. For an in-depth discussion of these principles, rules, and roles of medical care, refer to Field Manual (FM) 4-02.

1-2. The AHS is comprised of 10 medical functions. They are—
- Medical command and control.
- Medical treatment (organic and area support).
- Hospitalization.
- Medical evacuation (to include medical regulating).
- Dental services.
- Preventive medicine services.
- Combat and operational stress control.
- Veterinary services.
- Medical logistics (to include blood management).
- Medical laboratory support (to include both clinical laboratories and area laboratories).

1-3. Army Health System resources are arrayed across the battlefield in successive increment levels of increased capabilities (referred to as roles of medical care). Medical evacuation and the provision of en route medical care ensures an uninterrupted continuum of care is maintained while Soldiers are moved through the roles of medical care to the MTF best suited to treat the specific injuries.

SECTION II — MEDICAL EVACUATION

1-4. ***Medical evacuation*** **is the timely and effective movement of the wounded, injured, or ill to and between medical treatment facilities on dedicated and properly marked medical platforms with en route care provided by medical personnel.**

Chapter 1

1-5. The provision of en route care on medically equipped vehicles or aircraft enhances the patient's potential for survival and recovery and may reduce long-term disability.

- ***En route care*** **is the care required to maintain the phased treatment initiated prior to evacuation and the sustainment of the patient's medical condition during evacuation.**
- ***Patient movement*** **is the act of moving a sick, injured, wounded, or other person to obtain medical and/or dental treatment.**
- A *patient* is a sick, injured or wounded Soldier who receives medical care or treatment from medically trained personnel. (FM 4-02).

1-6. Medical evacuation support is provided on a direct support and area support basis. *Direct support* is a support relationship requiring a force to support another specific force and authorizing it to answer directly to the supported force's request for assistance. (FM 3-0). *Area support* is a method of logistics, medical support, and personnel services in which support relationships are determined by the location of the units requiring support. Sustainment units provide support to units located in or passing through their assigned areas. (ATP 4-90).

1-7. Medical evacuation supports the theater evacuation policy. The ***theater evacuation policy*** **is a command decision indicating the length in days of the maximum period of noneffectiveness that patients may be held within the command for treatment, and the medical determination of patients that cannot return to duty status within the period prescribed requiring evacuation by the first available means, provided the travel involved will not aggravate their disabilities or medical condition.**

1-8. Medical evacuation is an integral part of medical regulating. *Medical regulating* is the actions and coordination necessary to arrange for the movement of patients through the roles of care and to match patients with a medical treatment facility that has the necessary health service support capabilities and available bed space. (JP 4-02). (See Chapter 6 for more information on medical regulating.)

1-9. An efficient and effective MEDEVAC system—

- Minimizes mortality by rapidly and efficiently moving the sick, injured, and wounded to and between MTFs.
- Ensures continuum of care between roles of care.
- Serves as a force multiplier as it clears the battlefield using dedicated medical assets that enable the tactical commander to continue his mission with all available combat assets.
- Builds the morale of Soldiers by demonstrating that care is quickly available if they are wounded.
- Provides en route medical care that is critical in increasing survival, improving recovery, and reducing disability of the wounded, injured, or ill.
- Provides economy of force.
- Provides connectivity of the AHS as appropriate to the Military Health System.

The Army Health System and Medical Evacuation

PRIMARY TASKS

1-10. Table 1-1 discusses the primary tasks of MEDEVAC.

Table 1-1. Primary tasks and purposes of medical evacuation

Primary task	Purpose
Acquire and locate	Provide a rapid response to acquire wounded, injured, and ill personnel. Clear the battlefield of casualties and facilitate and enhance the tactical commander's freedom of movement and maneuver. This task is performed by the medical evacuation crew of the evacuation platform.
Treat and Stabilize	Maintain or improve the patient's medical condition during transport and provide en route care as required. This task is performed by medical evacuation crewmembers and providers when necessary.
Intratheater Medical Evacuation	Provide rapid evacuation utilizing dedicated assets to the most appropriate role of care. Provide a capability to cross-level patients within the theater hospitals and to transport patients being evacuated out of theater to staging facility prior to departure. This task is performed by the evacuation platforms in the medical company (ground ambulance) and medical company (air ambulance).
Emergency movement of medical personnel, equipment, and supplies	Provide a rapid response for the emergency movement of scarce medical resources throughout an operational environment.

1-11. The AHS is established in roles of increasing capability from the POI to definitive care. On the battlefield, casualties are evacuated rearward from point of injury or one role of care to the next higher role of care; the sequencing of this movement is dependent upon mission, enemy, terrain and weather, troops and support available, time available, and civil considerations (METT-TC) factors. In a contiguous battlefield, well established lines of communication, large numbers of casualties, and a wide array of MTFs can result in very deliberate evacuation from one sequential role to the next higher. In some situations, such as a noncontiguous battlefield, the allocation and availability of medical and MEDEVAC resources and the number, type, and criticality of patients being evacuated may require bypassing lower roles of care in order to ensure the timely treatment and appropriate care of casualties. The evacuation plan will be established by the appropriate level of command in coordination with the command's surgeon to ensure the best treatment is provided to all casualties.

ATTRIBUTES

1-12. The overall success of the Army's MEDEVAC system can be accredited to several attributes. These attributes include MEDEVAC assets that are dedicated resources that provide en route medical care, maintain continuity of care between roles of care, maintain connectivity to other AHS functions, provide economy of force and provide the commander with a force multiplier.

DEDICATED RESOURCES

1-13. The Army's MEDEVAC system is comprised of dedicated air and ground evacuation platforms. These platforms have been designed, manned, and equipped to provide en route medical care to patients being evacuated and are exclusively employed to support the medical mission. The focus of the MEDEVAC mission coupled with the dedicated platforms permit a rapid response to calls for support. The dedicated nature of this mission dictates that Army MEDEVAC units posture themselves ready to rapidly respond to evacuation missions and are not diverted to perform any other task. Medical evacuation resources are exclusively employed to support the medical mission and are therefore protected under the provisions of the Geneva Conventions from intentional attack by the enemy. (For a discussion of the Geneva Conventions in relation to air and ground MEDEVAC operations, refer to Appendix A of this publication and FM 4-02.)

EN ROUTE CARE

1-14. En route care is provided on all Army MEDEVAC platforms when a medical attendant is on board with access to the patient. This care is essential for minimizing mortality, enhancing survival rates, and reducing disability of wounded, injured, or ill Soldiers. Refer to paragraphs 1-16 through 1-24 for a discussion on the differences between MEDEVAC and operations.

1-15. The appropriate level of care must be maintained throughout the continuum of care. A patient who has received complex care requires continuous maintenance of the critical care support that was initiated at a forward MTF. Depending on the level of care, the medical personnel providing en route care may be critical care flight paramedics, en route critical care nurses, or other properly trained medical specialists. When possible, this en route care should be used as far forward as the METT-TC allows.

CONNECTIVITY

1-16. The synchronized employment of MEDEVAC resources provides and maintains the seamless continuum of care from the POI through successive roles of essential care within the theater. In addition to evacuating patients and providing en route medical care, MEDEVAC resources provide for the emergency movement of scarce medical resources such as critical Class VIII, blood, medical personnel, and medical equipment. Further, MEDEVAC resources are used to transfer patients from one MTF to another within the theater, to facilitate specialty care, as well as transferring patients from an MTF to an en route patient staging facility to enable intertheater evacuation. In addition to providing connectivity within the AHS, Army MEDEVAC resources provide MEDEVAC support and interface with MTFs of the other Services deployed in the theater. In joint and multinational operations, Army assets may provide this support and connectivity to joint and multinational forces within the joint operations area.

ECONOMY OF FORCE AND FORCE MULTIPLIER

1-17. Medical evacuation supports economy of force for low density, high demand medical specialties (such as a neurosurgeon), medical supplies, and medical equipment (such as computer tomography scans or magnetic resonance imaging equipment) by providing the capability to move the patient to the required care over long distances and by permitting the cross-leveling of medical supplies which reduces the need for large Class VIII stockpiles. By providing economy of force, the deployed medical footprint is reduced without negatively impacting on the care provided to the Soldier.

MEDICAL EVACUATION TOOLS

1-18. In addition to the attributes listed above, the Army's MEDEVAC system uses several tools to provide continuum of care by sustaining proximity between supported populations and MTFs. These tools enable patients to be efficiently evacuated and as a result, assists the maneuver commander to sustain mobility.

CASUALTY COLLECTION POINT

1-19. The ***casualty collection point*** **is a location that may or may not be staffed, where casualties are assembled for evacuation to a medical treatment facility.** Casualty collection points (CCPs) are normally predesignated along the axis of advance or evacuation routes. Forward of the battalion aid station (BAS), the combat medic, combat lifesaver, and combat troops take casualties to the CCPs. These points facilitate acquisition of casualties by supporting ambulance teams and reduce evacuation time. When used by the BAS, CCPs help preserve BAS mobility, preclude carrying casualties forward, and reduce evacuation time to the sustainment area.

1-20. When designating a CCP, the designating authority makes a decision whether or not to provide medical staff at the location. This decision is based upon the assessment of risk versus the availability of personnel. Normally, the role of care designating the point is responsible for staffing. Medical personnel may not be available to staff these points, and combat lifesavers and ambulatory patients may be required to perform self-aid, buddy aid, or enhanced first aid. The CCPs should be identified on operational overlays and planned by phase for operations. The CCP planning considerations include site security, proximity to the landing zone (LZ), cover and concealment, and access to evacuation routes. Leaders should address aid and litter teams, distribution of casualty equipment, and how choke points can be mitigated.

AMBULANCE EXCHANGE POINT

1-21. *Ambulance exchange point* is defined as **a location where a patient is transferred from one ambulance to another en route to a medical treatment facility.**

1-22. These ambulance exchange points (AXPs) are normally preplanned and are a part of the HSS appendix to the sustainment annex to the operation plan (OPLAN). In the forward area, the threat of enemy ground activities, large concentrations of lethal weapons systems, and effective use of antiaircraft weapons may dictate that the AXP be a predetermined rendezvous point for the rapid transfer of patients from one evacuation platform to another. The location of AXPs should be frequently changed to preclude attracting enemy fires.

1-23. Ambulance exchange points are established for many different reasons. For example, the ambulance platoon of the armored medical company (brigade support battalion) (BSB) possesses a mixture of wheeled and tracked ambulances, AXPs could be established for the following reasons:
- The brigade's tracked ambulances are provided so that they may keep up with maneuver elements. These vehicles carry the patients from the BAS to an AXP where the brigade wheeled ambulances take over for the relatively longer trip to the rear.
- Ambulance exchange points are not limited to ground evacuation assets. Another example is a situation where the threat air defense artillery capability is such that air ambulances cannot fly as far forward as the BASs. However, an AXP could be established a few kilometers to the rear, still well forward of the brigade support area.
- The brigade combat team (BCT) tracked or wheeled ambulances could then transfer the patients to the air assets, thereby facilitating the rapid evacuation of patients and realizing a significant timesavings.

1-24. By using AXPs, evacuation assets are returned to their supporting positions faster. This facilitates evacuation as the returning crews are familiar with the road network and the supported unit's tactical situation. In the case of air ambulance assets, AXPs are important because of the requirements for integration into the airspace control system at each role and the enhancement to survivability provided by current threat and friendly air defense information.

AMBULANCE SHUTTLE SYSTEM

1-25. The *ambulance shuttle system* is **a system consisting of one or more ambulance loading points, relay points, and when necessary, ambulance control points, all echeloned forward from the principal group of ambulances, the company location, or basic relay points as tactically required.** This system is an effective and flexible method of employing ambulances during operations. This system includes—
- *Ambulance loading point:* **This is the point in the shuttle system where one or more ambulances are stationed ready to receive patients for evacuation.**
- *Ambulance relay point:* **This is a point in the shuttle system where one or more empty ambulances are stationed to advance to a loading point or to the next relay post to replace departed ambulances.**
- *Ambulance control point:* **A manned traffic regulating, often stationed at a crossroad or road junction, where ambulances are directed to one of two or more directions to reach loading points and medical treatment facilities.** The need for control points is dictated by the situation. Generally, ambulance control points are more necessary in forward areas.

1-26. In the establishment of the ambulance shuttle system, once the relay points are designated, the required number of ambulances are stationed at each point. If the tactical situation permits, the ambulances may be delivered to the relay points by convoy.

1-27. Advantages of the ambulance shuttle system are that the system—
- Places ambulances at CCPs and BASs as needed.

Chapter 1

- Permits a steady flow of patients through the system to MTFs.
- Avoids unnecessary massing of transport in forward areas.
- Minimizes the danger of damage to ambulances by the enemy.
- Permits the commander or platoon leader to control their elements and enables them to extend their activities without advancing the headquarters.
- Facilitates administration and maintenance.
- Maximizes the use of small command elements (sections or platoons) to operate the ambulance shuttle without employing the entire parent unit.
- Provides for flexible use of other ambulance assets for specific situations.

STAFFING OF RELAY, LOADING, AND AMBULANCE CONTROL POINTS

1-28. Important points may be manned to supervise the blanket, litter, and patient movement items (PMI) exchange and to ensure that messages and medical supplies to be forwarded are expedited. Staffing also helps ensure information is passed along to ambulance crews.

SECTION III — MEDICAL EVACUATION VERSUS CASUALTY EVACUATION

1-29. There are critical differences in capabilities between MEDEVAC and casualty evacuation (CASEVAC). Planners and leaders must understand these differences in order to optimize conformity and continuity of medical care to reduce the mortality and disability of patients.

1-30. Medical evacuation is performed by dedicated, medically equipped, and standardized MEDEVAC platforms designed especially for the MEDEVAC mission to provide en route care by trained medical professionals who provide the timely, efficient movement and en route care of the wounded, injured, or ill persons from the battlefield or other locations to MTFs. The provision of en route care on medically equipped vehicles or aircraft greatly enhances the patient's potential for recovery and may reduce long-term disability by maintaining the patient's medical condition in a more stable manner.

1-31. Medical evacuation ground/air ambulance platforms are defined as: platforms exclusively employed for the evacuation and en route care of wounded, injured, or ill casualties and for the transport of medical personnel and equipment by military medical personnel.

1-32. The gaining MTF in coordination with the losing MTF is responsible for arranging for the evacuation of patients from the lower role of care. For example, Role 2 medical units are responsible for evacuating patients from Role 1 MTFs.

CASUALTY EVACUATION

1-33. A *Casualty* is any person who is lost to the organization by having been declared dead, duty status—whereabouts unknown, missing, ill, or injured. (JP 4-02). *Casualty evacuation* is the movement of casualties aboard nonmedical vehicles or aircraft without en route medical care. (FM 4-02). For more information on CASEVAC, refer to ATP 4-25.13.

> **WARNING**
>
> Casualties transported in this manner may not receive proper en route medical care or be transported to the appropriate MTF to address the patient's medical condition. If the casualty's medical condition deteriorates during transport, or the casualty is not transported to the appropriate MTF, an adverse impact on his prognosis and long-term disability or death may result.

1-34. If MEDEVAC platforms (ground or air) are available, casualties should be evacuated on these conveyances to ensure they receive proper en route medical care and reduced risk of attack due to legal protections afforded to properly marked medical evacuation platforms.

1-35. Since CASEVAC operations can reduce combat power and degrade the efficiency of the AHS, units should only use CASEVAC to move Soldiers with less severe injuries when MEDEVAC assets are overwhelmed or otherwise unavailable. Medical planners should ensure CASEVAC operations are addressed in the OPLAN/operation order (OPORD) as a separate operation, as these operations require preplanning, coordination, synchronization, and rehearsals. The CASEVAC plan should ensure casualties with severe or life-threatening injuries are prioritized for evacuation on dedicated MEDEVAC platforms.

1-36. When possible, nonmedical vehicles/aircraft transporting casualties should be augmented with a combat medic or combat lifesaver. The type of en route monitoring and medical care/first aid provided is limited by the following factors:
- Skill level of the individual providing care. (The combat medic is military occupational specialty-qualified [68W] to provide emergency medical treatment basic level care; the combat lifesaver is trained to provide enhanced first aid.) The combat medic can provide emergency medical intervention, whereas the combat lifesaver can only monitor the casualty and ensure that the basic lifesaving first aid tasks are accomplished.
- Medical equipment available.
- Number of casualties being transported.
- Accessibility of casualties—if nonstandard evacuation vehicle is loaded with the maximum number of casualties, the combat medic or combat lifesaver may not be able to attend to the casualties while the vehicle is moving. A combat medic may not have adequate space required to perform some interventions. If the condition of a casualty deteriorates and emergency measures are required, the vehicle will have to be stopped to permit care to be given.

1-37. Casualty evacuation and MEDEVAC are complimentary capabilities, and when used efficiently and effectively reduce Soldier mortality. Having CASEVAC capable platforms does not negate the need for planning for and using organic MEDEVAC assets. As complimentary capabilities, they enhance the maneuver commander's options and ability to clear their wounded from the engagement area, while ensuring that the more severely wounded have access to the increased lifesaving capabilities provided in the MEDEVAC platform.

MORTUARY AFFAIRS

1-38. At the unit level, commanders are responsible for the evacuation of human remains of assigned and attached personnel (military, DOD, civilian and contractor) to the nearest mortuary affairs facility. The movement of remains is an important logistical function but is not a task supported by MEDEVAC units or teams.

1-39. The evacuation of remains on MEDEVAC vehicles should be avoided due to a number of reasons such as—
- The MEDEVAC vehicles are a low density asset and must be responsive to the supported population.
- As MEDEVAC vehicle transports patients to MTFs, they must return quickly to continue or be prepared to conduct MEDEVAC operations.
- Adverse psychological impact to patients on MEDEVAC vehicles.

This page intentionally left blank.

Chapter 2
Army Medical Evacuation

Medical evacuation operations are planned to provide comprehensive, responsive, flexible, and agile support to the tactical commander to conform to the commander's intent and OPLANs. This chapter discusses the employment of MEDEVAC resources and the coordination and synchronization required to effectively execute MEDEVAC operations (to include the transfer of patients between MTFs and to En Route Patient Staging System (ERPSS) facilities by air and ground evacuation assets.

SECTION I — MEDICAL EVACUATION SUPPORT

2-1. Casualties requiring evacuation are prioritized to ensure the most seriously injured or ill receive timely medical intervention consistent with their medical condition. As with medical treatment, the patient's medical condition is the only factor used to determine the evacuation precedence. (Appendix A provides an in-depth discussion of the provisions of the Geneva Conventions and the law of war as they apply to MEDEVAC operations.)

2-2. The decision to request a MEDEVAC and the level of evacuation precedence will be made by the senior medical personnel on scene, or senior military ranking officer if medical personnel are unavailable based on the patient's condition and the tactical situation. Assignment of a MEDEVAC precedence is necessary. The precedence provides the supporting medical unit and controlling headquarters with information that is used in determining priorities for committing their evacuation assets. For this reason, correct assignment of precedence cannot be overemphasized; over classification may result in an increase in evacuation which could burden the AHS.

2-3. The patient's medical condition is the overriding factor in determining the evacuation platform and destination facility. The air ambulance operates wherever needed on the battlefield, dependent on risk and METT-TC factors. Air ambulances are staffed with critical care flight paramedics and a suite of medical equipment to provide optimum en route patient care including emergency medical intervention when required. It is the preferred method of evacuation for most categories of patients. Air ambulances are a low-density, high demand resource and must be managed accordingly. To conserve these valuable resources, medical planners should plan to use air ambulances to primarily move Priority I, URGENT and Priority IA, URGENT-SURG patients with the other categories on a space-available basis. Army Health System support planners must plan for a synchronized air and ground evacuation plan. Depending on the length of time required for an air ambulance to be dispatched and arrive at the POI, it may be prudent to evacuate the casualty by ground evacuation assets to a BAS or Role 2 MTF for stabilization by a physician. Commanders and planners ensure air and ground evacuation assets are positioned as close to the supported units as the tactical situation permits in order to evacuate Priority I, URGENT and Priority IA, URGENT-SURG patients to the appropriate role of care within one hour. Patients will be transported as soon as possible, consistent with available resources and pending missions. Table 2-1 on page 2-2 depicts the categories of evacuation precedence and the criteria used to determine the appropriate precedence.

Chapter 2

Table 2-1. Categories of evacuation precedence

Priority I—URGENT	Is assigned to emergency cases that should be evacuated as soon as possible and within a maximum of one hour in order to save life, limb, or eyesight and to prevent complications of serious illness and to avoid permanent disability.
Priority IA—URGENT-SURG	Is assigned to patients that should be evacuated as soon as possible and within a maximum of one hour who must receive far forward surgical intervention to save life, limb, or eyesight and stabilize for further evacuation.
Priority II—PRIORITY	Is assigned to sick and wounded personnel requiring prompt medical care. This precedence is used when the individual should be evacuated within four hours or if his medical condition could deteriorate to such a degree that he will become an URGENT precedence, or whose requirements for special treatment are not available locally, or who will suffer unnecessary pain or disability.
Priority III—ROUTINE	Is assigned to sick and wounded personnel requiring evacuation but whose condition is not expected to deteriorate significantly. The sick and wounded in this category should be evacuated within 24 hours.
Priority IV—CONVENIENCE	Is assigned to patients for whom evacuation by medical vehicle is a matter of medical convenience rather than necessity.
The NATO STANAG 3204 has deleted the category of Priority IV—CONVENIENCE, however, this category is still included in the United States Army evacuation priorities as there is a requirement for it in an operational environment.	

SECTION II — MEDICAL EVACUATION REQUESTS

2-4. This section discusses requests for MEDEVAC. The 9-line MEDEVAC request provides a standardized message format that helps expedite the medical evacuation process. The same format is used for both air and ground MEDEVAC requests. Specific procedures, frequencies, and security requirements for transmittal of MEDEVAC requests are delineated through the orders process and are made a part of the unit/command standard operating procedures (SOPs). Procedures for requesting MEDEVAC support must be institutionalized down to the lowest level. The 9-line MEDEVAC request format is shown in Appendix C on pages C-1 thru C-3.

2-5. The 9-line MEDEVAC request should be transmitted using secure communications for operational security. A technically capable enemy may be able to intercept a nonsecure 9-line MEDEVAC request and strike at the pickup zone causing additional harm to the patients and the MEDEVAC crew. Mission variables (METT-TC) must be taken into consideration, but the safety of a patient should not be further jeopardized if the 9-line MEDEVAC request cannot be sent by secure communications.

2-6. Each sector based on the METT-TC may be designated with a different method of evacuation as the primary means to effect evacuation. Sectors which have a high ground-to-air or air-to-air threat may rely on ground evacuation assets to move the majority of patients. In other sectors where the ground threat is high and comprised of small arms and explosive hazards (mines, improvised explosive devices, and unexploded ordinance), MEDEVAC operations may be more efficiently and effectively executed by air ambulances. An additional consideration in planning MEDEVAC operations is to determine whether armed escorts are required for either the ground or air ambulance mission. Those missions that require armed escorts must be thoroughly coordinated and synchronized between the medical assets and force protection assets that will accompany them. It may take a combination of both air and ground working in concert to mitigate the risk and to perform the evacuation through a chain of coordinated ground and air AXPs.

2-7. Medical evacuation requests often are sent from the POI, through intermediaries, such as higher headquarters, who then transmit the request up to the nearest MEDEVAC unit. The unit relaying the request must ensure that it relays the exact information originally received. The radio call sign and frequency relayed (Line 2 of the request) should be that of the requesting unit and not that of the relaying unit. However, the intermediaries contact information can be given as additional information if a callback is necessary to clarify details of the mission. Figure 2-1 on page 2-4 depicts examples of the communication flow for a MEDEVAC request.

Army Medical Evacuation

> **This subparagraph implements NATO STANAG 2087.**

2-8. In all circumstances, along with the patient's condition, the operational situation, terrain, weather conditions, enemy threat and availability of assets are considered when determining whether to send a ground or air ambulance.

2-9. Theater commanders may determine that additional information is required when submitting a MEDEVAC request. The major concern on adding additional requirements to a MEDEVAC request is that the addition does not delay the evacuation mission. Theater commanders should consult with their staff surgeon section to assess the requirement to add additional information. The additional information beyond the original format of the 9-line MEDEVAC request should be based off the medical benefit it provides to the condition of the patient being evacuated.

2-10. Some multinational partners may require/request additional information on the MEDEVAC request. This additional information may be included in a medical evacuation request to United States (U.S.) medical evacuation units. An example of this information would be the incorporation of the mechanism of injury, injury type, signs, treatment (MIST) report into the MEDEVAC request.

2-11. The MIST information is additional information and is sent as soon as possible after the 9-line MEDEVAC request has been sent. Medical evacuation missions should not be delayed while waiting for the MIST information.

2-12. The surgeons of the brigade, division, and combined joint task forces (CJTFs) must integrate air and ground MEDEVAC operations with the assigned combat aviation brigade (CAB). Successful MEDEVAC mission coordination requires the integration of aviation and medical functions as well as a decentralized approval process. Each air ambulance mission within the area of operations (AO) requires medical mission authorization and mission launch approval.

2-13. Medical mission authority begins at the theater-level through the creation of the theater evacuation policy and the medical rules of eligibility by appropriate medical officers. Once approved, these documents are published through the orders process and become the foundation for what constitutes a valid medical mission. For aeromedical evacuation missions, the medical approval authority is accomplished by verifying the details of the 9-line MEDEVAC request with the policy contained in the theater evacuation policy and/or medical rules of eligibility. Once confirmed that the mission request falls within the established theater guidance, the request becomes an approved medical mission. The validation of the medical necessity to generate a requirement can include—
- Transport of a patient.
- Patient precedence.
- Requirement for blood or blood products.
- Emergency resupply of medical related supplies, equipment, or personnel.

2-14. For aeromedical evacuation requests, the aviation commander considers the collective risk assessment of the mission and determines final execution or launch authority. The operational aspects related to the collective risk assessment include, but are not limited to—
- Patient care requirements.
- Threat of enemy action.
- Rules of engagement.
- Weather.
- Fighter management.
- Escort requirements.
- Tactical situation.

Chapter 2

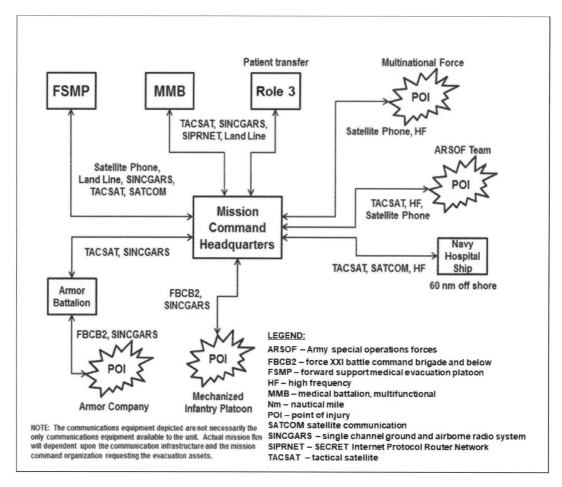

Figure 2-1. Flow of communication for medical evacuation request

GROUND EVACUATION REQUEST

2-15. Figure 2-2 provides an example of a request from the POI for ground evacuation. The patient is a PRIORITY precedence patient requiring evacuation from the POI/CCP to the supporting BAS (Role 1 MTF) with follow-on evacuation to an AXP for further evacuation to the medical company (brigade support battalion) (BSMC) (Role 2 MTF). If during evacuation the patient's medical condition deteriorates and their MEDEVAC precedence increases, an air ambulance asset could be requested. If the patient requires evacuation to a Role 3 MTF it can be accomplished by either an air or ground ambulance (GA) depending on the evacuation precedence and tactical situation.

Army Medical Evacuation

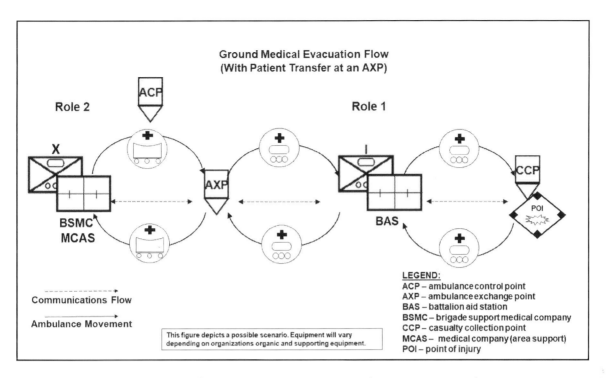

Figure 2-2. Ground evacuation request in a maneuver unit

2-16. Ground MEDEVAC assets also provide direct support and support on an area basis to units in the echelons above brigade (EAB). Figure 2-3 on page 2-6 depicts a medical company (area support) (MCAS) and the area support role in the EAB area of operations. The MCAS is replacing the area support medical company and the term MCAS is used interchangeably for both organizations in this publication. The *area of operations* is an operational area defined by a commander for land and maritime forces that should be large enough to accomplish their missions and protect their forces. (JP 3-0).

2-17. In the EAB area, the MCAS provides Roles 1 and 2 medical support on an area support basis. The MCAS has organic evacuation assets and receives augmentation when required from the medical company (ground ambulance) assigned to the medical battalion (multifunctional) (MMB).

2-18. Medical evacuation resources are also used to transfer a patient between MTFs within the AO and from an MTF to an en route patient staging system facility for further evacuation out of theater. In Figure 2-3, this mission is being accomplished by a ground ambulance. However, depending upon the distance from the MTF to the aeromedical staging facilities and/or aerial port of debarkation, this mission can also be accomplished by air ambulance assets.

Chapter 2

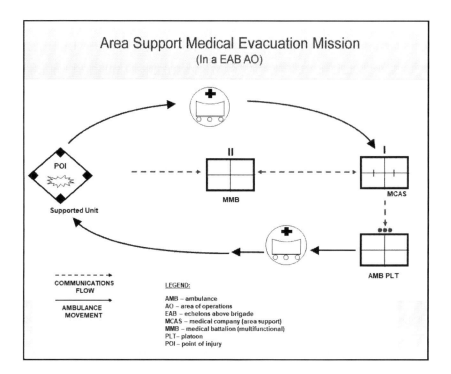

Figure 2-3. Ground evacuation request in the echelon above brigade

2-19. Air ambulances may fly as far forward as possible on the battlefield. Although evacuation by air ambulance is the preferred means for all casualties, when high evacuation workloads exist, evacuation by air ambulance should be the primary means used for URGENT and URGENT-SURG patients. The general support aviation battalion (GSAB)/CAB in coordination with the medical brigade, will position air ambulance assets where they can best support the tactical commander's plan through the timely and responsive evacuation. An informal medical operations cell may be established as a collaborative effort by the CAB or GSAB to assist in the coordination and synchronization with the supported ground commander's air and ground MEDEVAC plans. The staffing of this medical operations cell will be dependent on availability and location of medical planners from the CAB and GSAB staff. The medical operations cell will—

- Establish flight procedures specific to aeromedical evacuation missions within the CAB AO. This may include special routes or corridors as well as procedures for escort aircraft linkup.
- Ensure lines of communication to supported units and higher levels of medical command and control are available. Further, the medical operations cell ensures supported units understand MEDEVAC procedures and capability.
- Establish MEDEVAC briefing and launch procedures. Ensure there is 24-hour access to those able to launch high and very high-risk missions. See Army Regulations (AR) 40-3 and AR 95-1 for further information on launch authority.
- Maintain awareness of tactical and medical situation. Coordinate with medical regulating officer (MROs) at higher levels to efficiently conduct the general support mission and to work in concert with adjacent units.
- Assist the air ambulance commander and GSAB/CAB staff to conduct MEDEVAC operations.
- Coordinate missions with supported command surgeons to ensure situational understanding and coordination of MEDEVAC efforts, this staff keeps command surgeons informed on aeromedical evacuation missions being executed in their AO.
- Consult and coordinate with supported command surgeons when air ambulances cannot be launched to execute a requested mission. Ensures that the appropriate medical authority is notified to ensure mission can be accomplished by ground evacuation assets.

Army Medical Evacuation

- Perform intelligence tasking of air ambulance assets within the command or AO when required. For example, in an immature theater or during major combat operations, the communication capabilities of the Patient Evacuation Coordination Cell may be incapable or ineffective in performing intelligence tasking of air ambulances. In these situations, the medical operations cell of the CAB or GSAB may perform these duties in close coordination with the medical brigade and adjacent CAB or GSAB medical operations cells.

2-20. Figure 2-4 depicts a request for air ambulance support. The scenario shows the request being sent to the forward support medical evacuation platoon (FSMP), the air ambulance company, and the GSAB. The routing of the request would be dependent on how our forces are arrayed. In a similar situation where ground evacuation may not be readily available or the tactical situation precludes its use the requesting unit could transmit their medical evacuation request directly to a supporting FSMP as a primary means for a MEDEVAC request or the air ambulance company as a secondary option with a third being sent directly to the GSAB.

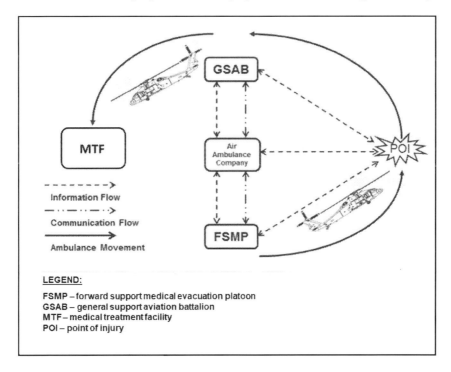

Figure 2-4. Aeromedical evacuation request

2-21. Figure 2-5 on page 2-8 illustrates evacuation zones in a noncontiguous AO. A noncontiguous AO is when one or more of a commander's subordinate forces' AOs do not share a common boundary. Examples of aeromedical evacuation support in a noncontiguous AO include—

- Forward support MEDEVAC platoons are placed where they are most needed. This can be with troops most often engaged in combat, high population density areas, areas of famine or disease with high civilian casualties, areas with dislocated persons, and geographically centralized locations.
- Area support MEDEVAC platoons or FSMPs accomplish the patient transfer mission that develops between FSMP and MTFs, between MTFs, and between MTFs and intertheater movement locations. These platoons also provide aeromedical evacuation support on an area basis in their immediate vicinity.

Chapter 2

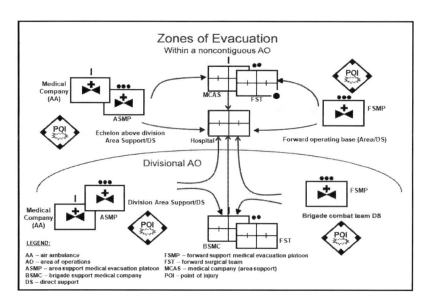

Figure 2-5. Air ambulance zones of evacuation

2-22. In contrast to a typical aviation mission cycle, continuous aeromedical evacuation coverage results in extended operational duty days that often exceed 24-hour period in length to provide seamless coverage to our forces. Medical evacuation units must plan and develop a detailed rhythm of military operations that addresses risk management, mission execution processes, and tactics, techniques, and procedures unique to 24-hour continuous evacuation operations.

2-23. A MEDEVAC crew cycle begins much like any other mission planning cycle; however certain required mission planning details will not be provided prior to the receipt of the 9-line MEDEVAC request. For instance, the crew begin their MEDEVAC planning cycle without the location, routes, threat at pick up point, patient precedence, number of patients, or time of execution for their mission. Without this critical information, the crew prepares their ambulance and medical equipment for the mission, and completes all known mission planning details. During a duty cycle it is critical that the crew responds as rapidly and as safely as possible when a mission is received. It is therefore imperative that crews are afforded the ability to manage rest while ensuring the procedures are in place to maintain battlefield situational understanding and facilitate a rapid, safe execution upon receipt of a MEDEVAC request. This includes but is not limited to crew rest cycles, mission and launch approval procedures, and aviation and logistical support to remote/split-based crews to ensure the safe and timely execution of missions.

JOINT AND MULTINATIONAL INTERCONNECTIVITY

2-24. When directed by the combatant commander (CCDR), Army MEDEVAC assets may be used to evacuate patients from the other Services or may be placed in direct support of other Services and/or multinational units participating in the operation. Communications and procedures should be synchronized prior to the initiation of the operation to ensure it is executed without preventable complications. Interoperability issues can result in slower response times and may adversely affect MEDEVAC operations. Figure 2-6 depicts a possible patient flow in a joint scenario, whereby Army MEDEVAC assets are supporting United States Marine Corps (USMC) operations and evacuating patients to the en route patient staging system facilities. When supporting multinational forces, U.S. military personnel may not be on site to provide security. Armed escorts and security forces may be required to accompany MEDEVAC missions involving multinational forces and this requirement should be included as a planning consideration so assets and personnel may be designated and integrated into training and rehearsals.

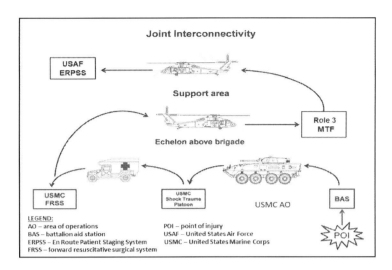

Figure 2-6. Air and ground ambulance evacuation in a joint environment

SECTION III — MEDICAL EVACUATION UNITS, ELEMENTS, AND PLATFORM CONSIDERATIONS

2-25. Evacuation platforms must be capable of keeping pace with the supported unit. The rapid evacuation of patients from POI to higher roles of care enables the maneuver formations to sustain initiative.

MEDICAL EVACUATION SUPPORT PROTOCOL

2-26. An effective and efficient medical evacuation system is performed by the higher role of medical care assets supporting forward and evacuating from the lower role or evacuation assets. This method of support ensures that far-forward evacuation elements are able to quickly deliver their patients and rapidly return to their formations, to maintain the ground commander's freedom of movement and maneuver. Medical evacuation support for EAB is provided by organic capabilities and as well as the MMB's medical company (ground ambulance), MCAS, and the GSAB's medical company (air ambulance).

2-27. The MMB maximizes the effective employment of ground ambulance resources whether supporting at EAB or theater level. It coordinates directly with the CAB's GSAB for effective use of air ambulance support with its MCASs ground ambulances. The medical brigade (support) (MEDBDE [SPT]), MMBs, GSAB, sustainment brigade, BCTs and division surgeon section coordinates and synchronizes MEDEVAC operations for the division AOs. The division surgeon section coordinates with the MEDBDE (SPT) for the evacuation of patients from division MTFs to supporting hospitals including the combat support hospital (CSHs) and its replacement, the hospital center currently being fielded. (For simplicity, the term "hospital" and "Role 3" will be used for both organizations in this publication.) The MEDBDE (SPT) coordinates and arranges the required intratheater MEDEVAC of patients to and from supporting hospitals. The medical command, deployment support (MEDCOM [DS]) coordinates for evacuation of patients out of the theater. To ensure that patients are evacuated to the appropriate MTFs, MROs and aeromedical evacuation officers (67J) are organic to the MEDCOM (DS), and MEDBDE (SPT).

MEDICAL EVACUATION PLATFORM FACTORS

2-28. Evacuation platforms must be capable of keeping pace with the troops supported. The patient's medical condition is the overriding factor in determining the evacuation platform and destination MTF. The air ambulance operates wherever needed on the battlefield, dependent on risk and METT-TC factors. Use of hardened armored MEDEVAC vehicles may be the vehicle of choice for some missions for short evacuation to an MTF or to a secure AXP for transfer to an air or wheeled ground ambulance. The criteria for an efficient and effective evacuation system were identified in Chapter 1.

Chapter 2

SECTION IV — MEDICAL EVACUATION AT UNIT LEVEL

2-29. The accumulation of casualties restricts unit movement and lowers morale. These influences are prevented through organized and timely evacuation from forward areas.

MEDICAL EVACUATION MISSION CONSIDERATIONS

2-30. The METT-TC factors affect all MEDEVAC missions and the employment of ambulance assets. The medical commander must consider the basic tenets that influence the employment of MEDEVAC assets. These factors include the patient's medical condition and the—

- Tactical commander's plan for employment of operational forces.
- Enemy's most likely course of action.
- Anticipated patient load.
- Expected areas of patient density.
- Availability of MEDEVAC resources to include ground and air crews.
- Availability, location, and type of supporting MTFs.
- Adherence to the protections afforded to medical personnel, patients, medical units, and medical transports under the provisions of the Geneva Conventions.
- Airspace control plan.
- Obstacle plans.
- Fire support plan (to ensure MEDEVAC assets are not dispatched onto routes and at the times affected by the fire support mission).
- Road network/dedicated MEDEVAC routes (contaminated and clean).
- Weather conditions.

PATIENT ACQUISITION

2-31. Units with organic MEDEVAC assets have the primary responsibility for patient acquisition. Methods of employment and evacuation techniques differ depending upon the nature of the operation.

2-32. Units without organic ambulance assets are provided MEDEVAC support on an area support basis. Units must develop techniques which facilitate the effective employment of their combat medics, enhance the ability to acquire patients in forward areas, and rapidly request MEDEVAC support. The techniques developed should be included in the unit SOP. At a minimum, the SOP should include the—

- Vehicle assignment for the organic medical personnel.
- Vehicles designated to be used for casualty transport and/or casualty evacuation.
- Procedures for requesting MEDEVAC support (during routine operations or during mass casualty situations). *Mass casualty* any number of human casualties produced across a period of time that exceeds available medical support capabilities. (JP 4-02).
- Role of the first sergeant, platoon sergeants, and combat lifesavers in MEDEVAC.

MANEUVER BATTALION MEDICAL PLATOONS

2-33. The medical platoon leader along with the medical platoon field medical assistant should be included in all battalion tactical planning. These officers—

- Maintain knowledge of the concept of operations, commander's intent, and the anticipated medical requirements.
- Develop the HSS/FHP plan (see ATP 4-02.55 and ATP 4-02.3) and provide HSS/FHP overlays with preplanned evacuation routes, CCPs and AXPs, to ambulance squads or teams for inclusion in the battalion OPLAN.
- Ensure that the medical platoon sergeant provides strip maps or other navigational tools to the ambulance drivers, if needed.

- Request augmentation support from the supporting medical company in advance of the operation, if required. When elements of other maneuver battalions are attached as part of a task force, they ensure that adequate medical elements are included in the support package.
- Ensure that orientation and support are provided for the medical personnel.
- Ensure that the use of air ambulances is coordinated into the medical support plan. This includes coordination with area support medical evacuation platoon (ASMP) or FSMP to establish and identify AXPs within evacuation routes that are suitable for aircraft for ground to air patient exchange.

2-34. The ambulance/evacuation section squad leader/noncommissioned officer ensures that the ambulance teams have a working knowledge of the terrain features in the AO. Whenever possible, they familiarize themselves with primary and secondary MEDEVAC routes through route reconnaissance conducted by the squad leaders. This platoon sergeant and squad leader manages the employment of the ambulance teams and monitors the communications net to remain abreast of the tactical situation.

2-35. The following factors should be considered when selecting ambulance routes:
- Tactical mission.
- Coordinating evacuation plans and operations with the unit movement officer.
- Security of routes and security escort.
- Availability of routes.
- Physical characteristics of roads and cross-country routes (to include natural obstacles).
- Requirements to traverse roads in urban areas and potential obstructions from rubble and debris.
- Traffic density.
- Time and distance factors.
- Proximity of possible routes to areas that may be subject to enemy fire.
- **Lines of patient drift** which are the natural routes along which wounded Soldiers may be expected to go back for medical care from a combat position.
- Cover, concealment, and available defilade for moving and stationary vehicles.
- Obstacle plans.
- Fire support plan (to ensure MEDEVAC assets are not dispatched onto routes and at the times affected by the fire support mission).

2-36. Depending upon the operational situation, the modes of evacuation may include walking Soldiers who are wounded, manual and litter carries, nonmedical CASEVAC assets, or dedicated MEDEVAC platforms. Evacuation in the battalion area normally depends on the organic ambulances assigned. Evacuation by air ambulance is dependent upon the patient's medical condition, availability of air assets, tactical situation, and weather conditions.

2-37. The ambulance team is normally deployed to each maneuver company trains and one remains with the Role 1 MTF/BAS in the combat trains. Ambulances operate as far forward as the tactical situation permits. An ambulance team operating in a maneuver company's AO, is normally under the tactical control of the maneuver company executive officer (XO) or first sergeant. This ambulance team, however, remains under the operational control (OPCON) of the medical platoon.

2-38. During the evacuation process, a casualty's equipment (weapons, ammunition, explosives, pyrotechnics, communications security equipment, and night vision device) is considered unit equipment and usually left with the unit. Individual protective equipment (body armor, helmet, and muzzles for military working dogs) if worn, is evacuated with the patient. Disease and nonbattle injury patients can be evacuated with other personal equipment (sleeping bag, personnel hygiene items) if space is available or if required by theater evacuation policy.

2-39. The medical operations officer ensures that the ambulances are located close to the anticipated patient workload. An ambulance team consists of one ambulance and two combat medics (on track vehicles, a third emergency care specialist is required to provide en route medical care). One or two of these teams serve in direct support of a maneuver company. Maneuver companies establish CCPs to facilitate the triage and evacuation of casualties. To become familiar with the specific terrain and battlefield situation, the team maintains contact with the supported maneuver company during most combat operations. Maneuver medical platoon ambulances not in direct support are positioned strategically throughout the battalion area or are sited at the BAS to—

- Evacuate patients from the company aid posts or CCPs to the BAS.
- Reinforce the forward positioned ambulance teams.
- Support the operating forces held in reserve or scout, mortar, and forward observer platoons.

2-40. Another employment option is to forward site the additional ambulance teams at company CCPs, as well as at the BASs. Many times the ambulance team finds casualties who have not been seen by a combat medic. In these cases, the team members dismount and then find, treat, and evacuate the patients. Ambulance teams not specifically dedicated to support combat elements can be used as messengers in medical channels and to provide transport of emergency medical personnel, equipment, and supplies.

2-41. During static situations where the maneuver company is not in enemy contact or is in reserve, the ambulance team returns to the BAS to serve as reinforcement to other elements in contact. However, during movement to contact, the ambulance team immediately deploys with its supported unit. In moving patients back to the CCP, the team may be assisted by nonmedical personnel. Specific duties of the ambulance team are to—

- Maintain contact with supported elements.
- Find and collect the wounded.
- Administer tactical combat casualty care (TCCC).
- Initiate or complete the Department of Defense (DD) Form 1380 (Tactical Combat Casualty Care Card). For instructions on completing DD Form 1380, see References section of this publication (Referenced Forms).
- Evacuate patients to the BAS.
- Direct or guide ambulatory patients to the BAS.
- Resupply company/platoon medics.
- Serve as messengers in medical channels.
- Perform route reconnaissance from Role 1 MTF to Role 2 MTF or the higher role of medical care.

AREA MEDICAL EVACUATION SUPPORT

2-42. Medical evacuation on an area basis is required at all roles in the AHS system. The EAB units without organic MEDEVAC resources, such as combat engineers, will require evacuation support on an area basis. Air ambulance FSMP have an area support responsibility even when in a direct support role to another organization. This allows units that are transitioning through the FSMPs AO without organic evacuation assets to have coverage. To ensure that these elements receive adequate support, the medical planner must include these requirements into the OPLAN. Prior coordination is essential to ensure that the locations of CCPs, AXPs, and BASs are disseminated to these elements and that any unique support requirements are included in the OPLAN. The triggers for relocating these MEDEVAC assets to their next locations should also be included.

MEDICAL EVACUATION SUPPORT FROM ROLE 1 AND ROLE 2 MEDICAL TREATMENT FACILITIES

2-43. Evacuation from Role 1 MTFs operated by a treatment squad or team is normally provided by the BSMC/MCAS ambulance platoon and the GSAB air ambulance company. These ambulance assets provide MEDEVAC support on an area basis to other units in the brigade area, as required.

GROUND AMBULANCES

2-44. The ambulance teams from the ambulance platoon are normally collocated with the BSMC/MCAS treatment platoon for mutual support. They establish contact and locate one ambulance team with the medical platoon of each maneuver battalion. The remaining ambulances are used for BCT operations and area support. The ambulances may be positioned at AXPs or CCPs, or are field-sited with Role 1 or Role 2.

Army Medical Evacuation

AIR AMBULANCES

2-45. A FSMP from the supporting GSAB's medical company (air ambulances) may provide direct support air ambulance support for the BSMC Role 2 MTF. The FSMP normally remains OPCON to the GSAB, and is staged forward to support the BCTs. When a direct support relationship is defined, the direct support relationship provides authority to the BSMC (in coordination with the brigade surgeon's section) to direct the integrated air and ground MEDEVAC support for the BCT. The platoon leader of the FSMP should be included in the brigade tactical planning process.

2-46. The air ambulance crew evacuates Priority I, URGENT patients from as far forward as possible to the BSMC. Further, when a forward surgical team (FST) or forward resuscitative surgical team (FRST) is collocated with a BSMC, air ambulances evacuate Priority IA, URGENT-SURG to this facility. It may be necessary to skip roles of care when a patient's condition would benefit from going directly to a Role 3 MTF and the tactical situation permits.

2-47. The GSAB air ambulances may operate from the support areas providing around-the-clock, immediate response, evacuation aircraft, as well as provide the BSMC commander flexibility and agility in the emergency movement of treatment teams and medical equipment to the forward battle area. It also provides emergency movement of Class VIII, blood, and blood products. External lift capabilities of MEDEVAC helicopters add an important dimension to their role on the battlefield in moving medical supplies and equipment.

BRIGADE MEDICAL EVACUATION PLAN

2-48. The brigade surgeon section is responsible for developing the BCT MEDEVAC plan. The BSMC commander provides synchronization for the execution of brigade MEDEVAC plan, to include the use of both ground and air assets. The brigade surgeon section medical planner should include the BSMC commander, XO, brigade personnel staff officer, brigade XO, medical platoon leaders, FSMP leader, the brigade support battalion medical plans and operations officer, and the support operations officer in the planning process.

2-49. In the BSMC, the XO is the principal assistant to the commander for the tactical employment of the company assets. The XO should be included in all brigade tactical planning. Officer preparation will be required to reinforce or reconstitute forward HSS elements and to request augmentation through the brigade surgeon section, if required. The medical evacuation platoon leaders keep the BSMC XO apprised of the unit's operational capability and readiness. This enables the XO to effect timely reinforcement or augmentation. The BSMC XO must be familiar with the specific terrain and battlefield situation. Further, the XO should have a thorough understanding of the division and brigade commanders' ground tactical plan.

GROUND AMBULANCE EVACUATION SUPPORT TO BRIGADE ROLE 2 MEDICAL TREATMENT FACILITIES

2-50. Evacuation from the BSMC is provided by ground ambulances from the supporting ground ambulance company, MCAS, or from air ambulances from the GSAB medical company (air ambulance). The medical companies (ground ambulance) provide support to the BSMCs and medical elements in the EAB. They usually evacuate only those patients who will not return to duty within 72 hours.

GROUND AMBULANCE SUPPORT FOR ECHELONS ABOVE BRIGADE UNITS

2-51. The MCAS ambulance platoon normally collocates with the treatment platoon for mutual support and area taskings. It performs ground evacuation and en route patient care for supported units in the EAB. The MCAS may also evacuate patients from the supported Role 2 MTFs in the division support area to the MCAS Role 2 MTF. The ambulance platoon is mobile in operations as its assets may be totally deployed at one time. When not fully deployed, the ambulance platoon normally forward stations a portion of its teams in support of those units in the division support area.

2-52. The remaining teams are used for general support to the EAB where medical organic assets are not assigned and for reinforcing support or ambulance shuttles. Platoons or squads from the medical company (ground ambulance) will be in direct support, or OPCON to, and collocated with supported EAB assets.

SECTION V — EXCHANGE OF PROPERTY

2-53. United States Army MEDEVAC operations require that whenever a patient is evacuated from one MTF to another, or is transferred from one ambulance to another, medical items of equipment (CASEVAC bags [cold weather type bags], blankets, litters, and splints) remain with the patient.

2-54. To prevent rapid and unnecessary depletion of supplies and equipment, the receiving Army element exchanges like property with the transferring element. This reciprocal procedure will be practiced to the fullest extent possible through all phases of evacuation from the most forward element through the most rearward hospital.

2-55. With United States Air Force (USAF) aeromedical evacuation operations, a major factor in the evacuation of patients is that specific medical equipment and durable supplies designated as PMI must be available to support the patient during the evacuation. Examples of PMI include—
- Ventilators.
- Patient monitors.
- Pulse oximeters.

2-56. These items will be available for exchange at the supporting aeromedical staging facilities. Refer to Joint Publication (JP) 4-02 and Army techniques publication (ATP) 4-02.1 for additional information on PMI.

This paragraph implements NATO STANAG 2128 and ABCA Standard 436.

2-57. Medical property of multinational nations with ratified standardization agreements (STANAGs) North Atlantic Treaty Organization (NATO) and American, British, Canadian, Australian, and New Zealand armies (ABCA) accompanying multinational patients will be returned to the parent nation at once, if possible. If it is not possible, like items will be exchanged as in the paragraph above.

2-58. When medical property of multinational forces without ratified standardization agreements accompanies patients of multinational forces, it is returned as soon as practical. Commanders should consult with their staff judge advocate early in the planning process to ensure appropriate policy and procedures are developed and disseminated.

SECTION VI — KEY ACTIVITIES DURING SPECIFIC OPERATIONS

2-59. The primary tasks and attributes of MEDEVAC operations are consistent throughout all operations. The following paragraphs provide examples of activities that may be conducted by MEDEVAC units, by specific operation. Figures for MEDEVAC support during these operations are located in Appendix D.

KEY ACTIVITIES DURING OPERATIONS TO SHAPE

2-60. Operations to shape consist of various long-term military engagements, security cooperation, and deterrence missions, tasks, and actions intended to assure friends, build partner capacity and capability, and promote regional stability (FM 3-0). These operations occur across the joint phase model and focus on four purposes—
- Recognizing and countering adversary attempts to gain positions of relative advantage.
- Promoting and protecting U.S. national interests and influence.
- Building partner capacity and partnerships.
- Setting conditions to win future conflicts.

2-61. Medical evacuation units will continue to train on individual, collective, mission essential task list (commonly known as METL) tasks, conduct clinical training and rotations to sustain medical skills and certifications, and pursue professional development courses. Medical evacuation units must be proficient not only to conduct the primary MEDEVAC tasks, but to rapidly disassemble their equipment, move to a new location, and re-establish themselves to respond to MEDEVAC requests, under all weather conditions.

Medical planners should attend staff courses such as the joint medical planner's course when available, especially for 67J and 70H (officer) and 68W (enlisted) occupational specialties on division or corps staff.

2-62. Staff exercises may be held at tactical through operational levels to train and rehearse the planning and operation of a medical evacuation system including CCPs, AXPs, evacuation routes, MEDEVAC requests, synchronization of MTFs, and medical regulating. Units may prepare time-phased and deployment data, update equipment sets, and prepare containers and vehicles for deployment.

2-63. Key shaping activities may include support to military exercises and involve additional requirements such as support to ship to shore and shore to ship or overwater missions. Associated requirements such as deck-landing qualification, Dunker and Helicopter Emergency Egress Device System (HEEDS) (breathing apparatus) training should be identified early in order to ensure it is included in training plans.

2-64. Military engagements include interactions with foreign military medical personnel and foreign and domestic civilian authorities. Medical evacuation units can support security cooperation and security force assistance goals through activities such as medical evacuation training to build partner capacity. Medical evacuation support to foreign internal defense may be constrained by the number of medical evacuation vehicles and type and location of MTFs. In some instances, CASEVAC may be utilized to move a casualty to a MTF or an AXP manned by MEDEVAC aircraft due to extended distances and limited assets. Humanitarian assistance missions provide Army MEDEVAC units opportunity to perform its mission while strengthening partnership between the U.S. and the supported nation.

KEY ACTIVITIES DURING OPERATIONS TO PREVENT

2-65. The purpose of operations to prevent is to deter adversary actions contrary to U.S. interests (FM 3-0). This may require intervention of U.S. forces to restore stability and will likely mean units will be called to conduct operations under severe time constraints.

2-66. At the strategic level, the theater army plans and coordinates Army capabilities to set the theater. Planners assess available intelligence to identify medical threats, foreign medical capabilities and infrastructure, population at risk, viable routes for evacuation and locations for MTFs and units. Planners may also develop contingency or operation plans. During operations to prevent, a corps headquarters may deploy into an operational area as a tactical headquarters with subordinate divisions and brigades as a show of force, or, may deploy an early entry command post to provide control over arriving forces. Medical planners within the corps surgeon section should begin planning early, develop an understanding of the mix of forces and how to best support them with the available medical evacuation assets which may include units and capabilities from other Services.

2-67. Units and personnel who are part of a readiness force or who have been designated for specific operation will have completed clinical rotations and training events and be available for immediate recall and deployment. During operations to prevent, units could be part of a tailored force in support of force projection. Force projection is the ability to project the military instrument of national power from the United States or another theater, in response to requirement for military operations (JP 3-0).

2-68. Medical evacuation units may provide area MEDEVAC support during reception, staging, onward movement and integration, at ports of embarkation, debarkation, and along movement routes. Medical planners may coordinate with U.S. or host nation organizations for some aspects of support.

KEY ACTIVITIES DURING LARGE-SCALE COMBAT OPERATIONS

2-69. Medical evacuation activities during large-scale combat operations are discussed in detail in subsequent sections of this chapter including offense, defense, air assault and airborne operations. Medical evacuation primary tasks remain consistent throughout large-scale combat operations.

2-70. During large-scale combat operations against a peer threat, units will simultaneously conduct actions to seize, retain, and exploit the initiative. The complexity and lethality of the environment will require MEDEVAC units to operate across multiple domains (air, land, maritime, space, and cyberspace), in a synchronized effort with the MTFs to clear the battlefield thereby sustaining the initiative of the maneuver commander.

2-71. A key aspect of large-scale combat operations is its joint nature. Medical planners and commanders can mitigate problems by providing a plan that synchronizes MEDEVAC, CASEVAC, and treatment capabilities, addresses constraints and limitations, and standardizes terms and procedures. Other Service representatives should be included into planning efforts when feasible and integrated into battle rhythm events pertaining to MEDEVAC and MTFs.

KEY ACTIVITIES DURING OPERATIONS TO CONSOLIDATE GAINS

2-72. Consolidate gains are the activities to make enduring any temporary operational success and set the conditions for a stable environment allowing for the transition to control to legitimate authorities Army doctrine reference publication. It occurs within sections of an area of operations where large-scale combat operations are no longer occurring. Consolidation of gains consists of security and stability tasks but may include combat against remnant or bypassed enemy forces. This presents a threat to MEDEVAC units and teams utilizing evacuation routes or manning AXPs and should be mitigated by accompanying security assets.

2-73. Medical evacuation support during the consolidation of gains may be on an area or direct support basis to maneuver forces conducting consolidation of gains in corps or division consolidation areas. Medical evacuation operations follows the traditional support provided to combat forces.

2-74. Medical evacuation support during operations to consolidate gains may require a MEDEVAC company to provide direct support to maneuver forces in one area while supporting stability tasks in another. A key factor for MEDEVAC support will be for the host nation to reestablish its own ability to provide medical services for its population to a reasonable level it possessed prior to hostilities and to support the legitimacy of the host nation. Additional information on MEDEVAC support in operations dominated by stability tasks can be found in Section VII of this chapter.

SECTION VII — MEDICAL EVACUATION SUPPORT FOR OPERATIONS CONDUCTING OFFENSE, DEFENSE, AND STABILITY TASKS

2-75. This section addresses MEDEVAC support for operations conducting offense, defense and stability tasks.

MEDICAL EVACUATION SUPPORT FOR THE OFFENSE

2-76. A number of factors such as the proliferation of advanced technologies and the modernization and professionalization of peer and near-peer competitors has increased the lethality and likelihood of offensive operations (see FM 3-0). The ability to conduct air or ground MEDEVAC may be significantly limited during large-scale combat operations against near-peer competitors. United States forces directly engaged in ground combat may be unable to provide additional resources in support of MEDEVAC due to the intensity of battle and need to maintain initiative. Offensive operations require close synchronization of MEDEVAC and CASEVAC assets due to the likelihood of higher casualty rates.

2-77. The main purpose of the offense is to defeat, destroy, or neutralize the enemy force. The main feature of the offense is taking the initiative. The aim of the commander in the offense is to expedite the outcome. Audacity, concentration, surprise and tempo characterize the conduct of offensive tasks and are components of initiative. Initiative, combined with maneuver, makes possible the conduct of decisive offensive tasks Army doctrine publication (see ADP 3-90).

2-78. There are four primary offensive tasks that are conducted to defeat and destroy enemy forces and seize terrain, resources and population centers. The four primary offensive tasks are movement to contact, attack, exploitation, and pursuit.

2-79. When considering the evacuation plans to support an offensive action, the medical planner must consider many factors. The forms of maneuver (envelopment, frontal attack, infiltration, penetration, and turning movement), as well as the enemy's capabilities, influence the character of the patient workload and its time and space distribution. The analysis of this workload determines the allocation of AHS resources and the location or relocation of MTFs. Evacuation support of offensive operations must be responsive to several essential characteristics. As operations achieve success, the areas of casualty density move away from the supporting facilities. This causes the routes of MEDEVAC to lengthen. Heaviest patient workloads occur during disruption of enemy main defenses, at terrain or tactical barriers, during the assault on final objectives, and during enemy counterattacks. The accurate prediction of these workload points by the medical planner is essential if MEDEVAC operations are to be successful.

2-80. In offensive operations, the major casualty area is normally the zone of the main attack. As the main attack accomplishes the primary task of the division, it receives first priority in the allocation of combat power. The allocation of combat forces dictates roughly the areas which are likely to have the greatest casualty density. As a general rule, all BCT MTFs are located initially as far forward as operations permit. This allows the maximum use of these facilities before lengthening evacuation lines force their displacement forward. In deep operations with chemical, biological, radiological, and nuclear (CBRN) weapons targeted at supporting logistical bases, mass casualty operations may be conducted in sustainment areas.

2-81. As advancing operational formations extend control of the battle area forward, supporting medical elements overtake patients. This facilitates the acquisition of the battle wounded and reduces the vital time elapsed between wounding and treatment. In offensive operations, two basic problems confront the supporting evacuation units—

- Contact with the supported unit must be maintained. Responsibility for the contact follows the normal pattern of rear to front. The contact is maintained by forward deployed air and ground evacuation resources.
- Mobility of the MTFs supporting the operational formations must be maintained. Periodically, BCT medical companies and collocated FSTs and FRSTs are cleared of their patients so that they may move forward. In some scenarios, Role 3 hospitals may also have to evacuate patients to reduce patient overflow, surgical backlogs, or to increase bed availability.

2-82. This requirement for prompt evacuation of patients from forward MTFs requires available ambulances to be leveled well forward from the outset. The requirement for periodic movement of large numbers of patients from MTFs in the EAB areas further stresses the evacuation system.

2-83. Medical evacuation planning for offensive operations has many of the same considerations for each offensive task. Additional planning factors are—

- Patient evacuation may be slow and difficult due to damage to roads or inaccessibility of patients.
- Ground evacuation routes may be unsecured.
- Evacuation distance may increase as the operation continues.
- Medical evacuation will have increased reliance on convoys and air ambulances.
- Ambulances can be echeloned well forward at each role of care to quickly evacuate the patients generated by suddenly occurring contact.
- Medical treatment facilities, including BASs and treatment teams, moving with their respective formations may overtake patients and thereby reduce delays in evacuation and treatment.
- When the isolated nature of the envelopment maneuver precludes prompt evacuation, the patients are carried forward with the treatment element.
- During river crossing operations, maximum use of air ambulance assets is made to prevent excessive patient build-up on far shore. Near shore MTFs are placed as far forward as assault operations and protective considerations permit to reduce ambulance shuttle distances from off-loading points.

AIRBORNE AND AIR ASSAULT OPERATIONS

2-84. Airborne and air assault operation are some of the most complicated military operations to undertake, not only from an operational perspective, but also from a sustainment and AHS perspective. Planning for these operations requires coordination with organic and supporting medical units and the aviation units that control the air ambulances. For additional information on airborne and air assault operations refer to FM 3-99.

2-85. During the initial phases of airborne and air assault operations medical planners should plan for casualty movement by way of manual carries, opportunity lift, and designated CASEVAC assets. Until

Chapter 2

evacuation assets are available, units must be prepared to provide medical care through their attached and organic medical assets.

AIRBORNE OPERATIONS

2-86. An airborne operation involves the air movement into an objective area of combat forces and their logistic support for execution of a tactical, operational, or strategic mission. The means employed may be any combination of airborne units, air transportable units, and types of transport aircraft, depending on the mission and the overall situation.

2-87. Medical evacuation of casualties from a drop zone or airhead is difficult until the area achieves a level of security and MEDEVAC means are available; the BCT must be prepared to provide medical care through its organic medical company or the attachment of EAB medical elements. The evacuation of patients will be by litter bearers or by organic vehicles when they arrive, back to the CCP/BAS. Further evacuation to higher roles of care may be delayed until the lodgment is secure enough to permit rotary or fixed wing MEDEVAC or CASEVAC vehicles to land or, until a maneuver ground unit reaches the area. The initial evacuation from the airhead may be CASEVAC after the first aircraft arrives to deliver reinforcements and ammunition.

AIR ASSAULT OPERATIONS

2-88. An air assault operation is an operation in which assault forces, using the mobility of rotary-wing assets and the total integration of available firepower, maneuver under the control of a ground or air maneuver commander to engage enemy forces or to seize and hold key terrain.

2-89. In air assault operations, the evacuation is by litter bearers to CCP/BAS and then by air ambulance to a Role 2 MTF. Evacuation to a Role 2 MTF may initially be provided as opportunity by CASEVAC backhaul due to aerial sequencing, threat, or logistical prioritization.

2-90. The CAB allocates MEDEVAC assets to the supported air assault task force for the duration of the air assault. However, the size and distance of the planned air assault dictates the duration of MEDEVAC support to the air assault task force. As a general rule, the supporting commander should provide MEDEVAC assets to the supported commander until ground lines of communication are established.

2-91. The evacuation platoon leader from the BSMC and the air ambulance platoon leader from the air ambulance company conduct the MEDEVAC planning for the air assaults in coordination with the air assault task force and supported unit's staff.

2-92. When planning for MEDEVAC support for air assault operations—
- Integrate ground evacuation measures into the overall MEDEVAC plan.
- Plan MEDEVAC routes to supporting Roles 2 and 3 MTFs. Ensure routes are briefed to all aircrews participating in the air assault.
- Plan for medical personnel to fly on CASEVAC aircraft if they are available and the time and the situation permit.
- Ensure MEDEVAC crews are available for air assault orders, rehearsals, and preparations.
- Brief CCP and AXP locations during the air assault rehearsal.
- Plan to maintain a forward arming and refueling point after the air assault is completed so that MEDEVAC aircraft have a place to stage from for follow-on ground tactical operations.

2-93. Medical evacuation aircraft are limited assets and should be scheduled and used accordingly. The air assault task force's casualty estimate provides planning guidance for the number of MEDEVAC aircraft needed to support the air assaults. To maximize the number of mission hours required to support the mission, MEDEVAC aircraft should be staged to support an air assault at the latest possible time. Medical evacuation aircraft should support short distance air assaults from the pickup zone or brigade support area. To expedite pick up of casualties in long distance air assaults, aircraft may stage at a forward arming and refueling point or use a restricted operating zone.

2-94. During the execution of an air assault operation, the air ambulances are controlled by the air mission commander. This allows for the deconfliction of airspace and the coordination of security escorts if required. The air mission commander also coordinates the backhaul of casualties on designated aircraft which return the casualties to the CCP at the pickup zone for further medical care (if medical personnel are available) and MEDEVAC to receiving MTFs.

MEDICAL EVACUATION SUPPORT FOR THE DEFENSE

2-95. The inherent strengths of the defense include the defender's ability to occupy positions before the attack and use the available time to prepare the defenses. Commanders choose to defend to create conditions for a counter offensive that allows Army forces to regain the initiative. Other reasons for conducting a defense include retaining decisive terrain or denying a vital area to the enemy, to attrite or fix the enemy as a prelude to the offense, in response to surprise action by the enemy, or to increase the enemy's vulnerability by forcing the enemy to concentrate forces.

2-96. A feature of the defense is to regain the initiative from the attacking enemy. The defending commander uses the characteristics of the defense—disruption, flexibility, mass and concentration, preparation, and security—to help accomplish that task.

2-97. There are three basic defensive tasks—area defense, mobile defense, and retrograde. These apply to both the tactical and operational levels of war, although the mobile defense is more often associated with the operational level. These three tasks have significantly different concepts and pose significantly different problems. For a more in-depth discussion of these defense tactics and control measures refer to ADP 3-90.

2-98. Support is generally more difficult to provide in the defense. The patient load reflects lower casualty rates, but forward area patient acquisition is complicated by enemy actions and the maneuver of combat forces.

2-99. Medical personnel generally are permitted much less time to reach the patient, complete vital tactical combat casualty care, and remove him from the POI. Increased casualties among exposed medical personnel further reduce the medical treatment and evacuation capabilities. Heaviest patient workloads, including those produced by enemy artillery and CBRN weapons, may be expected during the preparation or initial phase of the enemy attack and in the counterattack phase.

2-100. The enemy attack may disrupt ground and air routes and delay evacuation of patients to and from treatment elements. The depth and dispersion of the defense create significant time and distance problems for evacuation assets. Operational elements may be forced to withdraw while carrying their remaining patients to the rear. The enemy exercises the initiative early in the operation which may preclude accurate prediction of initial areas of casualty density. This makes the effective integration of air assets into the evacuation plan essential. The use of air ambulances must be coordinated and synchronized with the supporting GSAB to ensure the synchronized execution of air and ground evacuation operations occurs.

2-101. The support requirements for retrogrades may vary widely depending upon the tactical plan, the enemy reaction, and the METT-TC factors. Firm rules that apply equally to all types of retrograde operations are not feasible, but considerations include the following—

- Requirement for maximum security and secrecy in movement.
- Influence of dislocated person's movement that may impede MEDEVAC missions conducted in friendly territory.
- Integration of evacuation routes and obstacle plans.
- Difficulties in controlling and coordinating movements of the force which may produce lucrative targets for the enemy.
- Movements at night or during periods of limited visibility.
- Time and means available to remove patients from the battlefield. In stable situations and in the advance, time is important only as it affects the physical well-being of the wounded. In retrograde operations, time is more important. As available time decreases, medical managers at all roles closely evaluate the capability to collect, treat, and evacuate all patients.
- Medical evacuation routes will also be required for the movement of troops and materiel. This causes patient evacuation in retrograde movements to be more difficult than in any other type of operation. Difficulties include—
  - Mission command and communications may be disrupted by the enemy.

Chapter 2

- Successful MEDEVAC requires including ambulances on the priority list for movement.
- Providing for the transportation of the slightly wounded in cargo vehicles and providing guidance to subordinate commanders defining their responsibilities in collecting and evacuating patients.
- Special emphasis must be placed on the triage of patients and consideration given to the type of transportation assets available for evacuation.
● When the patient load exceeds the means to move them, the tactical commander must make the decision as to whether patients are to be left behind (abandoned). The medical staff officer keeps the tactical commander informed in order that he may make a timely decision. Medical personnel and supplies must be left with patients who cannot be evacuated.
● During a withdrawal air ambulance assets must be prepared to support the withdrawing force as well as forces left in contact. Planning considerations include—
- The medical company (air ambulance) element supporting the withdrawing force must ensure it maintains sufficient station capability at its present location to support the maneuver elements while taking action to move to a new location.
- Air ambulance elements will normally evacuate to MTFs in EAB sustainment area positions to avoid impeding the withdrawal of the BSMC within the brigades.
● The air ambulance company may conduct rapid movement of medical personnel and accompanying equipment and supplies to meet the requirements for mass casualty, reinforcement and reconstitution, or emergency situations to include expeditious delivery of whole blood, biological, and medical supplies to meet critical requirements.

MEDICAL EVACUATION SUPPORT FOR OPERATIONS DOMINATED BY STABILITY TASKS

2-102. Stability is an overarching term encompassing various military missions, tasks, and activities conducted outside the U.S. in coordination with other instruments of national power to maintain or reestablish a safe and secure environment, provide essential governmental services, emergency infrastructure reconstruction, and humanitarian relief. Stability tasks may be conducted alone or simultaneously during unified land operations along with offensive and defensive tasks.

2-103. Operations dominated by stability tasks promote and protect U.S. national interests by influencing the threat, political, and information dimensions of an operational environment. They include developmental, cooperative activities during peacetime and coercive actions in response to crisis. For more information on stability tasks see ADP 3-07.

2-104. Stability tasks may complement the main effort or they may themselves constitute the main effort. Stability tasks include; establishing civil security, conducting security cooperation, establishing civil control, restoring essential services, supporting governance, and supporting economic and infrastructure development.

2-105. Medical evacuation support to forces conducting stability tasks is dependent upon the specific type of operation, anticipated duration of the operation, number and location of forces deployed, theater evacuation policy, medical troop ceiling, and anticipated level of violence. In most situations, MEDEVAC support follows the traditional support provided to combat forces. If there is a shortened theater evacuation policy, a limited medical troop ceiling, and limited hospitalization assets within the AO, organic and direct support ambulance support is provided from the POI to the supporting Roles 1 or 2 MTF and, once the patient is stabilized for further evacuation, from the treatment element to an airfield for evacuation out of the theater.

SECTION VIII — MEDICAL EVACUATION SUPPORT FOR ARMY SPECIAL OPERATIONS FORCES

2-106. Army special operations forces (ARSOF) do not have an organic MEDEVAC system. The ARSOF is, therefore, dependent upon the conventional theater MEDEVAC system for this support. The ARSOF does have an organic capability to affect CASEVAC using ARSOF airframes (those used for infiltration/extraction of ARSOF personnel). During aerial CASEVAC, the casualty may not receive en route medical care unless specific planning and coordination is accomplished to staff the airframe with medically trained personnel prior to the execution of aerial CASEVAC operations.

2-107. Army special operations forces are those Active and Reserve component Army forces designated by the Secretary of Defense that are specifically organized, trained and equipped to conduct and support special operations. The acronym ARSOF represents Civil Affairs, Military Information Support operations, Rangers, Special Forces, Special Mission Units, and Army special operations aviation forces assigned to the United States Army Special Operations Command. Army Special operations forces are oftentimes required to operate in remote, austere, and denied environments with or without indigenous forces for prolonged periods of time without direct support from any military medical infrastructure, such as facilities, personnel, or MEDEVAC. Therefore, they must be self-sustaining in all areas of medical care throughout all spectrums of operations creating a situation where ARSOF medical personnel assume the responsibility and implementation for AHS support for ARSOF. Special Forces, Ranger and Special Operation Civil Affairs medical personnel receive enhanced medical training above that provided to conventional combat medics. Additionally, nonmedical ARSOF personnel may receive medical training at the combat lifesaver level providing for an enhanced first aid capability. For more information on AHS support to ARSOF refer to ATP 4-02.43. For more information on the Special Forces medical sergeant and the Special Operations combat medic refer to ADRP 3-05.

2-108. Since Special Forces and Rangers often operate far removed from conventional AHS, they must be more self-reliant and sustaining than conventional forces. The tactical combat casualty care initiative originated with the United States Army Special Operations Command. Special operations forces (SOF) do not have a dedicated, designed, and equipped MEDEVAC capability; therefore, their mission profile often requires the use of nonmedical platforms augmented with medical personnel to perform the evacuation function. The categories of MEDEVAC precedence are not changed, and both MEDEVAC and CASEVAC may occur depending upon the availability of assets and the time window available to execute the evacuation mission. Time is of the essence to remove the casualty as quickly as possible to where further treatment can be provided.

2-109. Ideally, MEDEVAC for ARSOF personnel should follow the doctrinal flow sequence. The ARSOF medical planner must be innovative and follow the tenets of immediate far forward stabilization. He directs evacuation to the appropriate MTF when the condition of the patients warrants it, with whatever means of transportation are available.

2-110. Medical evacuation of ARSOF casualties is an operational matter. That is, it must reflect the commander's concept of the operation. It can only succeed when the medical planner integrates the MEDEVAC plan with the tactical plan and logistics airflow.

2-111. Planning factors to consider when planning MEDEVAC support to ARSOF includes—
- Small unit and austere AHS capability. Army SOF will require support on an area basis. Army SOF unit location (geographical factors and time distance factors) may require collocation of assets.
- Remote operational areas and long evacuation routes. Army SOF units operate at a long distance from airfields suitable for evacuation. Army SOF often operate in areas that impede evacuation by rotary-wing aircraft or where aviation assets are not available. This places a premium on the early application of trauma management and casualty stabilization. Evacuation of ARSOF casualties may require the transfer of patients between multiple rotary wing aircraft.
- Medical evacuation, medical regulating and patient tracking requires an understanding of sensitivity of ARSOF missions. During evacuation, they must account for any sensitive items and documents that the ARSOF patient possesses.

2-112. Conventional MEDEVAC assets are not normally used when operations occur in hostile and denied areas. The conventional MEDEVAC system will normally be used once the casualty is extracted from the hostile or denied area. For more information of MEDEVAC planning and AHS for ARSOF refer to ATP 4-02.43.

SECTION IX — MEDICAL EVACUATION IN URBAN OPERATIONS

2-113. Military conflicts routinely take place in urbanized terrain. Urban operations are those military actions planned and conducted on a terrain where man-made structures impact on the tactical options available to the commander.

URBAN TERRAIN AND ENVIRONMENT

2-114. Urban terrain is characterized by a three-dimensional battlefield, having considerable rubble, ready-made fortified fighting positions, and an isolating effect on all operating forces. In this environment, the requirement for a sound and understandable evacuation plan cannot be overstated. Of concern to medical and tactical planners is the need to plan, train, prepare, and equip for evacuation from under, above, and at ground level. Refer to ATP 3-06/Marine Corps Techniques Publication (MCTP) 12-10B for more information on urban operations.

TERRAIN CONSIDERATIONS

2-115. Medical evacuation in the urban environment is a labor-intensive effort. Due to rubble, debris, barricades, and destroyed roadways, much of the evacuation effort will be by manual litter teams. When this occurs, establish a litter or ambulance shuttle system. The shuttle system reduces the distance that litter teams have to carry the wounded or injured Soldiers. This enhances the litter team's effectiveness by providing brief respites, reducing fatigue. Further, the litter teams stay in the forward areas. They are familiar with the geography of the area and what areas have or have not been searched for casualties. In moving patients by litter, you should—

- Use covered evacuation routes such as subways, whenever possible.
- Use easily identifiable points for navigation to CCPs.
- Rest frequently by alternating litter teams.

EVACUATION PLANNING CONSIDERATION FOR URBAN OPERATIONS

2-116. Conducting MEDEVAC operations in the urban environment challenges the medical planner. He must ensure that the AHS plan includes special or unique materiel requirements or improvised use of standard equipment. The plan must be sufficiently flexible to support unanticipated situations.

2-117. Special equipment requirements include, but are not limited to—

- Axes, crowbars, and other tools used to break through barriers.
- Special harnesses; portable block and tackle equipment; grappling hooks; poleless, nonrigid and collapsible litters; lightweight collapsible ladders; heavy gloves; and casualty blankets with shielding. This equipment, using pulleys, is for lowering casualties from buildings or moving them from one building to another at some distance above the ground.
- Equipment for the safe and quick retrieval from craters, basements, sewers, and subways. Casualties may have to be extracted from beneath rubble and debris.
- Air ambulances equipped with a rescue hoist may be able to evacuate casualties from the roofs of buildings or may be able to insert needed medical personnel and supplies.

> **DANGER**
>
> There is a possibility of a *dynamic rollout* occurring when proper tension is not maintained at all times during hoist operations. Excessive slack in the hoist cable may allow the hook and seat attachment point to become positioned such as to allow for *dynamic rollout*. For alternate attaching procedures for all affected equipment and for more information, refer to message dated 151200Z Aug 16, SUBJECT - SAFETY OF FLIGHT (SOF), TECHNICAL, RESCUE HOIST OPERATIONAL, RESTRICTIONS, H-47-16-SOF-04/H-60-16-SOF-02/H-72-16-SOF-01. (This information can be found at the web site listed in the references section of this publication.)

2-118. Communication may be limited within the urban environment. Medical evacuation teams will have difficulty in locating injured or wounded Soldiers due to their isolation within buildings or by being hidden in rubble and debris. Casualties may have to use visual forms of communication to mark their position, such as markers, panels, or field expedients (equipment or parts of uniforms) indicating where they may be found.

2-119. Due to the increased hazards to aviation assets operating in an urban environment, MEDEVAC from POI or CCP may be limited to ground ambulances. Ground ambulance may need to transfer patients to air ambulances at preplanned AXPs outside the build-up area that provides greater security and access.

ESTABLISHING CASUALTY COLLECTION POINTS

2-120. Preplan and establish CCPs at relatively secure areas accessible to both ground and air ambulances. The locations of these points are depicted on the HSS/FHP overlay to the OPLAN. Casualty collection points should—

- Offer protection from enemy fires.
- Be as far forward as the tactical situation permits.
- Be identifiable by an unmistakable feature (natural or man-made).
- Allow rapid turnaround of ambulances.

2-121. The tactical commander must approve route markings to the MTF and whether to display the Geneva Red Cross at the facility. (Camouflaging or not displaying the Geneva Red Cross can forfeit the protections, for both medical personnel and their patients afforded under the Geneva Conventions. Refer to Appendix A for additional information.) The location of the MTF must be as accessible as possible, but well separated from fuel and ammunition depots, motor pools, reserve forces, or other lucrative enemy targets, as well as civilian hazards such as gas stations or chemical factories.

GROUND EVACUATION FOR URBAN OPERATIONS

2-122. When using ground evacuation assets in support of urban operations, the medical planner must be aware that build-up areas may have significant obstacles to vehicular movement. Factors requiring consideration include—

- Rubble and other battle damages that complicate and canalize transportation operations within the urban terrain.
- Bypassed pockets of resistance and ambushes pose a constant threat along evacuation routes.
- Land navigation with most tactical maps proves to be difficult. Using commercial city maps when available can aid in establishing evacuation routes.
- Ambulance teams may have to dismount from the ambulance, search for, and rescue casualties.
- Movement of patients becomes a personnel intensive effort. There are insufficient medical personnel to search for, collect, and treat the wounded. Medical units may require assistance in the form of litter bearers and search teams from supported units.

Chapter 2

- Dislocated persons may hamper movement into and around urban areas.
- Dislocated civilians and detainees receive medical treatment according to rules of medical eligibility and the Geneva Conventions.

AEROMEDICAL EVACUATION FOR URBAN OPERATIONS

2-123. When using air ambulance assets in support of urban operations, the medical planner must consider enemy air defense capabilities and terrain features, both natural and man-made, within and adjacent to the build-up areas. Air ambulances may be the preferred means of evacuation in the AO, but special consideration must be taken into consideration when operating in urban areas. These considerations include the following:

- Telephone and electrical wire and communication antennas can hinder aircraft movement.
- Availability of secure landing zones.
- Landing zones may include buildings with helipads on their roofs or sturdy buildings, such as parking garages.
- Large open sports fields, parks, and promenades can present suitable LZs but must be generally free of telephone and electrical wires. Such open areas may be difficult for small maneuver elements to secure.
- Hoist operations on the top of small buildings or confined areas may be preferred but also exposes the aircraft and crew to potential hazards.
- Snipers with air defense capabilities may occupy upper stories of the urban area's taller buildings.

2-124. Medical personnel require special training in the tactics, techniques, and procedures required to operate in an urban environment. If they are to survive in this environment, they must know how to—

- Cross open areas safely.
- Avoid barricades and explosive hazards (mines, improvised explosive devices, and unexploded ordnance).
- Enter and depart safely from buildings.
- Recognize situations where booby traps or ambushes are likely and would be advantageous to the enemy. Detailed information on the conduct of combat operations in the urban environment is contained in ATP 3-06/MCTP 12-10B and ATP 3-06.1/Marine Corps Reference Publication (MCRP) 3-35.3A/Navy, tactics, techniques, and procedures (NTTP) 3-01.04/Air Force tactics, techniques, and procedures (AFTTP) 3-2.29

Note. Medical personnel do not engage in offensive-type actions. They must rely on the supported unit to provide covering fires and to clear rooms and buildings prior to entry.

SECTION X — ARMY SUPPORT TO CIVIL AUTHORITIES

2-125. Army forces support civil authorities by performing defense support of civil authorities (DSCA) tasks.

DEFENSE SUPPORT OF CIVIL AUTHORITIES

2-126. Defense support of civil authorities is defined as support provided by U.S. Federal military forces, DOD civilians, DOD contract personnel, DOD component assets, and National Guard forces (when the Secretary of Defense, in coordination with the Governors of the States, elects and requests to use those forces in Title 32, United States Code, National Guard) in response to requests for assistance from civil authorities for domestic emergencies, law enforcement support, and other domestic activities, or from qualifying entities for special events.

2-127. The primary purpose of all DSCA missions are to—
- Save lives.
- Alleviate suffering.
- Protect property.

2-128. Defense support of civil authorities encompasses support provided by the components of the Army to civil authorities within the U.S. and its territories. This includes support provided by Regular Army, Army Reserve, and the National Guard. Army forces operating within the U.S. encounter very different operations and environments than they face outside the Nation's boundaries. Principally, the roles of civilian organizations and the relationship of military forces to federal, state, tribal, and local agencies are different. The support provided by Army forces depends on specific circumstances dictated by law, as state and federal laws define almost every aspect of DSCA. The differences are pronounced enough to define a different task set than offense, defense, or stability. Soldiers and Army Civilians need to understand domestic environments so they can employ the Army's capabilities efficiently, effectively, and legally. For more information on support to civil authorities see ADP 3-28, ATP 4-02.42, and Department of Defense directive (DODD) 3025.18.

2-129. Although not the primary purpose for which the Army is organized, trained, and equipped, DSCA operations are a vital aspect of the Army's service to the Nation. The skills that allow Soldiers to accomplish their missions on today's battlefield can support local, state, and federal civil authorities, especially when domestic emergencies overwhelm the ability of government agencies to support fellow Americans.

2-130. The four primary characteristics that define DOD support are—
- State and federal laws define how military forces support civil authorities.
- Civil authorities are in charge and military forces support them.
- Military forces depart when civil authorities are able to continue without military support.
- Military forces must document costs of all direct and in direct support provided.

2-131. Homeland security and homeland defense are complementary components of the National Security Strategy (see references section of this publication for web site). In both homeland defense and homeland security, the Army may conduct DSCA.

HOMELAND SECURITY

2-132. The DOD supports the Nation's homeland security effort, which is led by the Department of Homeland Security. Homeland security is the concerted national effort to prevent terrorist attacks within the U.S.; reduce America's vulnerability to terrorism, major disasters, and other emergencies; and minimize the damage, and recover from attacks, major disasters, and other emergencies that occur.

2-133. To preserve the freedoms guaranteed by the Constitution of the U.S., the Nation must have a homeland that is secure from threats and violence, including terrorism. Homeland security is the Nation's first priority, and it requires a national effort in which the DOD has a key role in that effort. The National Strategy for Homeland Security complements the National Security Strategy of the U.S. by providing a comprehensive framework for organizing the efforts of federal, state, local, and private organizations whose primary functions are often unrelated to national security. Critical to understanding the overall relationship is an understanding of the distinction between the roles that DOD plays with respect to securing the Nation and homeland security and the policy in the National Strategy for Homeland Security, which has the Department of Homeland Security as the lead (see references section of this publication for web site). Homeland security at the national level has a specific focus on terrorist threats. The DOD focus in supporting homeland security is broader. Military application of the National Strategy for Homeland Security calls for preparation, detection, deterrence, prevention, defending, and responding to threats and aggression aimed at the homeland.

2-134. An understanding of the National Incident Management System is important to all response partners as this national crisis response system provides a consistent, nationwide approach to prepare for, respond to, and recovery from domestic emergencies, regardless of cause, size or complexity.

Chapter 2

2-135. Medical evacuation support plans to augment civil authorities may be necessary for a variety of contingencies. Defense support of civil authorities planning must address a range of problems such as—
- Early identification of MEDEVAC capabilities, units, and personnel available to support various contingencies large enough to require DSCA.
- Command and control relationships between civil authorities and DOD forces especially when DOD units are split-based.
- Cost capture and reimbursement from civil authorities to DOD in non-Stafford Act emergencies.
- Support for deployed DOD forces when no DOD logistics operations are deployed, including medical support.

2-136. During DSCA command of military forces remains within military channels, but missions begin as requests for assistance from the supported civil authorities. Additional considerations include communications compatibility, terminology usage, and legal constraints and how they may affect the mission. They must also identify who their counterparts are and who they will be interacting with on the local, state, and federal levels.

2-137. Medical evacuation assets may be deployed as part of a larger military force that would include medical or aviation organizations. Medical evacuation assets, both air and ground may provide services similar to what they do when supporting unified land operations. This support would include the evacuation of patients, movement of medical supplies and personnel, and support of search and rescue activities.

HOMELAND DEFENSE

2-138. Homeland defense is the protection of U.S. sovereignty, territory, domestic population, and critical defense infrastructure against external threats, and aggression or other threats as directed by the President. Missions are defined as homeland defense if the nation is under concerted attack from a foreign enemy. Department of Defense leads homeland defense and is supported by the other federal agencies. The purpose of homeland defense is to protect against and mitigate the impact of incursions or attacks on sovereign territory, the domestic population, and defense critical infrastructure. For homeland defense missions, as directed by the President of the U.S. or the Secretary of Defense, DOD serves as the lead federal agency.

2-139. Medical evacuation support to homeland defense operations may follow the same tasks as found in unified land operations. These MEDEVAC tasks can be conducted in the same manner and under the same constraints as they are during unified land operations. In support of homeland defense, MEDEVAC assets, both ground and air would perform patient evacuation and transfers, movement of medical supplies and personnel, and search and rescue mission support.

MEDICAL EVACUATION SUPPORT FOR FOREIGN HUMANITARIAN ASSISTANCE

2-140. Medical evacuation support to foreign humanitarian assistance and DSCA are very similar in their execution and planning. The largest difference between the two actions is where they are conducted and the regulations and laws that govern their execution.

2-141. Foreign humanitarian assistance operations may use DOD forces to assist foreign civil authorities as they prepare for or respond to crises and relieve suffering. In these operations, DOD forces provide essential support, services, assets, or specialized resources to help civil authorities deal with situations beyond their capabilities. The purpose of which is to meet the immediate needs of designated groups for a limited time, until those foreign civil authorities can do without DOD assistance.

2-142. Foreign humanitarian assistance operations can include a number of activities such as disaster relief, assistance to dislocated persons, the provision of medical care to isolated populations, and refeeding programs resulting from famines or natural disasters. Medical evacuation assets may be used to evacuate the injured from disaster sites, to provide the emergency transport of critically needed medical supplies and personnel to remote locations, or to perform emergency rescues during times of flooding, earthquakes, or other natural disasters.

SECTION XI — OTHER TYPES OF MEDICAL EVACUATION SUPPORT MISSIONS

2-143. Medical evacuation support includes less common mission types. Planners and leaders should be familiar with the variety of missions they will be required to conduct and support.

EVACUATION OF MILITARY WORKING DOGS

2-144. Injured or ill military working dogs may be evacuated on any transportation means available. The using unit is responsible for the evacuation of the animal. Use of dedicated MEDEVAC assets (air or ground ambulances) is authorized based on mission priority and availability. When possible, the handler should accompany the animal during the evacuation to ensure MEDEVAC personnel safety. Units requesting MEDEVAC for military working dogs should include the location of veterinary treatment facilities or support units in their request.

PERSONNEL RECOVERY OPERATIONS

2-145. Army personnel recovery is the sum of military, diplomatic, and civil efforts to prevent isolation incidents and to return isolated persons to safety or friendly control. Army personnel recovery operations are covered in FM 3-50.

2-146. Air ambulances do not conduct personnel recovery operations but may serve in a support role. Air ambulances flying in contested or denied areas are not protected from attack unless flying at times and routes agreed to by the enemy and may be summoned to land. Air ambulances must obey a summons to land. Personnel retain their protections under the Geneva Conventions. However, if air ambulances participate in the actual search and rescue phases of the operation, they are not solely engaged in the provision of AHS and are, therefore, not afforded the protections. Casualties may be transferred to an air ambulance at an exchange point located outside of contested or denied areas. Personnel recovery aircraft may be augmented with medical personnel and equipment for the mission and is the preferred method. If the involvement in these operations consists solely of evacuating wounded crewmembers from a crash site in friendly territory, air ambulances retain the protection accorded to them under the provisions of the Geneva Conventions.

SHORE-TO-SHIP EVACUATION OPERATIONS

2-147. Lessons learned from past operations have shown that U.S. Army helicopters should be able to operate to and from U.S. Navy air-capable ships. An inter-Service agreement between the Army and the Navy allows for deck-landing qualification of Army pilots (see FM 1-564 for additional information) and the current memorandum of understanding with the Navy. It is important that units having contingency missions requiring Navy support establish training requirements to obtain naval operations orientation, water egress training, water survival, proficiency in specific terminology used in Navy air to ship operations, and deck-landing qualification. This enhances the successful accomplishment of the MEDEVAC mission to naval vessels and mitigates risk.

2-148. The Military Sealift Command operates two hospital ships, the U.S. Naval Ship MERCY T-AH 19 and the U.S. Naval Ship COMFORT T-AH 20. One ship is based on each coast and, when needed, will be assigned medical staffs from military hospitals, getting underway within 5 days. The hospital ships MTFs were designed for a total patient capacity of 1,000, including 500 acute care beds and 500 recuperation beds. The hospital ships have 50 trauma stations in the casualty receiving area, 12 operating rooms; a 20-bed recovery room; 80 intensive care beds; and 16 intermediate, light, limited care wards. The maximum patient flow rate, for which the helicopter facility and the casualty reception area were designed, is 300 patients per 24 hours. There is a limited capability to receive casualties from boats.

2-149. The U.S. Army has the shore-to-ship MEDEVAC mission on an area support basis for Marine forces deployed on land.

Chapter 2

COMMUNICATIONS

2-150. In past joint operations, communications have been burdensome for both Army and Navy elements. It is essential to establish commonality of communication requirements during training exercises and to establish communication equipment and frequencies for MEDEVAC to Navy vessels. This will provide smooth integration of Army helicopters into the Navy airspace management system during actual operations.

NAVIGATION

2-151. As Navy vessels may operate relatively long distances from the ground combat operations, Army air ambulances units need to be proficient in over water navigation. The use of navigational aids from the Navy element in support of the operation is the first priority for over water navigation.

MEDICAL EVACUATION OF DETAINEES

2-152. Sick, injured, and wounded detainees are treated and evacuated in military police channels when possible. They must be physically segregated from U.S. and multinational patients. Providing guards for the transport of detainees is not the responsibility of MEDEVACs units or the MTF. Guards for these detainees are provided according to the BCT, division or corps orders and are from other than medical resources. The echelon commander is normally responsible for this support. The U.S. provides the same standard of medical care for wounded, sick, and injured detainees as that given to U.S. and multinational Soldiers. Wounded, sick, or injured detainees in the theater may be treated and returned to military police channels for evacuation, or the detainees may be stabilized and moved through medical channels to theater MTFs for treatment. Detainees are not evacuated from the theater for medical treatment.

2-153. When detainees are evacuated through medical channels, medical personnel—
- Report this action through medical channels to detainee operations medical director and the next higher headquarters.
- Request disposition instructions from the MEDBDE (SPT) patient movement branch.

2-154. The MEDBDE (SPT) patient movement branch is responsible for—
- Coordinating the transportation means.
- Identifying the MTF to which the detainees will be taken.
- Coordinating, in conjunction with the MTF commander, with the Detainee Reporting System to account for detainees within medical channels.

SECTION XII — MEDICAL EVACUATION IN SPECIFIC ENVIRONMENTS

2-155. Medical evacuation in specific environments or circumstances demand additional consideration. The MEDEVAC effort must be well planned and its execution synchronized to be effective. Further, MEDEVAC personnel must be flexible and ready to improvise, if needed, to meet the demands of unique situations.

MOUNTAIN OPERATIONS

2-156. In the past, armies have experienced great difficulty in evacuation of patients from mountainous areas because they are extremely diverse in nature. Some mountains are dry and barren with temperatures ranging from extreme heat in the summer to extreme cold in the winter. In tropical regions, mountains are frequently covered by lush jungles and heavy seasonal rains occur. Many areas display high rock crags with glaciated peak and year-round snow cover. Elevations can also vary from as little as 1,000 feet to over 16,000 feet above sea level with drastic and rapidly occurring weather changes. For more information on mountain operations, refer to ATP 3-90.97.

2-157. Operations in mountainous terrain require some procedure modifications. This is due to the environmental impact on personnel and equipment. Important physical characteristics and considerations that influence MEDEVAC are—
- Rugged peaks, steep ridges, and deep valleys.
- Limited number of trafficable roads.
- Reduced communication ranges.
- Unpredictability of and severe changes in weather.
- Decreased partial pressure of oxygen.
- Limited availability of LZs.

2-158. In order to effectively support the tactical plan, the AHS plan must provide maximum flexibility. The AHS planner should consider using all means of evacuation. Due to the length of evacuation times and the limited means of ground evacuation, it is important to triage and prioritize patients prior to movement. Planning considerations include—
- The availability of improved, hard-surfaced roads is extremely limited, if they exist at all. Usually, improved roads are only found in valley corridors. Such roads are often dependent upon a system of narrow bridges spanning mountains streams and ravines. They may also twist along ridgelines and cling to steep shoulders.
- Secondary roads and trails may be primitive and scarce. However, they may provide the only routes capable of vehicular traffic. Cross-compartment travel between adjacent valleys may be impossible by ground vehicle. Off-road travel requires detailed planning, even for short distances.
- Because of rough terrain, the Role 2 MTF may not be able to reach the BAS by ground vehicle. An ambulance shuttle system can be established with an AXP for air and ground evacuation vehicles to meet litter bearers. Litter bearers and beasts of burden such as donkeys or horses may be the only means of evacuation available. Any available personnel may be used as litter bearers (nonmedical personnel from supported units may be required to augment the litter bearer teams). Close coordination between Role 2 MTFs and BASs in establishing CCPs or AXPs is necessary to—
  - Reduce distance traveled by litter bearers.
  - Reduce evacuation time.
  - Conserve personnel.
  - Locate the best potential LZs for air ambulances.
- In mountainous areas, evacuation of patients by air is the preferred means. Air ambulances permit the rapid movement of patients over rugged terrain. For example, to travel a distance of only 6 kilometers on foot could take up to 2 hours, while flying time could be less than 2 minutes.
- If frequency modulated radios are the principal means of communication in this environment, the ability to transmit may be hampered by the limitations of line-of-sight-transmissions.
- The briefing of ambulance drivers needs to be extensive, including detailed strip maps and overlays. Further, specific instructions on what to do in various situations should be covered (such as if the vehicle breaks down or the unit moves).

2-159. The mountain environment, with its severe and rapidly changing weather, impacts on aircraft performance capabilities, accelerates crew fatigue, and requires special flying techniques. Having to rely on continuous aviation support for a successful mountain operation can be risky. Challenges and planning considerations to air ambulance operations in mountainous environment include—
- Lack of LZs is critical because the characteristics of mountain terrain do not usually afford adequate LZs. The terrain may only allow the aircraft to hover while loading patients onboard.
- Hoist operations can be expected in mountainous terrain requiring the use of the rescue hoist. When possible, orientation and training sessions with supported troops should be conducted to help minimize the difficulty of such missions.
- Enemy air defenses must be considered because when enemy air defense capabilities preclude using air ambulances in forward areas, they should be used to evacuate patients from AXPs or from Role 2 MTFs.

Chapter 2

- Ambulatory patients may be reported as litter patients in mountainous terrain because these patients may be unable to move unassisted over the rugged terrain. Once placed on the air ambulance, their status may be upgraded.
- Additional crew training for air ambulance crews should include mountain flying techniques and aeromedical factors. For ground MEDEVAC crews, training should include mountaineer skills as taught in the Army mountain warfare school and as prescribed in ATP 3-90.97.

2-160. Soldiers operating in mountainous areas are exposed to other injuries and illness that frequently occur in this environment. These conditions include—
- An increased rate of fracture, sprains, and dislocation injuries.
- Incidents of acute mountain sickness, high-altitude pulmonary edema, and cerebral edema caused by rapid ascent to heights over 7,500 feet above sea level.
- Cold weather illness.
- Dehydration and heat exhaustion.
- Sunburns and snow blindness.
- Aggravated sickle cell anemia. (Although this condition is not considered a mountain illness, personnel with the sickle cell trait can be seriously affected by the decrease in barometric pressure and lower oxygen levels found at higher altitudes.)

2-161. The proportion of litter cases to ambulatory cases is increased in mountainous terrain, for even the slightly wounded may be unable to move unassisted over rough terrain. Litter relay stations may be required along the evacuation route to conserve the energy of litter bearers and to speed evacuation. It is important to be able to predict the number of patients that can be evacuated with available personnel. When the average terrain grade exceeds 20 degrees, the four-man litter team is no longer efficient and should be replaced by a six-man team. The average mountain litter team should be capable of climbing 120 to 150 vertical meters of average mountain terrain and return with a patient in approximately 1 hour. More information on litter teams can be found in ATP 4-25.13.

JUNGLE OPERATIONS

2-162. Army health system support elements in a jungle environment retain the same basic capabilities as in other environments. Jungle operations, however, subject personnel and equipment to effects not found in other environments. The jungle environment degrades the ability to maneuver. Security problems are also increased and affect MEDEVAC operations as much as they do the combat force. For more information of jungle operations refer to FM 90-5.

2-163. In jungle operations, combinations of air and ground evacuation units are used to maximize the patient evacuation potential. Using this dual system of evacuation ensures that the inherent limitations of one system can be compensated for by the other. Jungle variations affect the organizing, positioning, and securing of AHS assets. Due to the terrain, aerial resupply is usually a common practice. The responsiveness provided by aerial resupply requires fewer supplies to be stockpiled in the combat trains.

2-164. Jungle combat operations are usually characterized by ambushes and unconventional warfare-type operations. The security threat caused by infiltrators requires that lines of communication be patrolled often and that convoys be escorted. It is, therefore, essential that AHS be performed as far forward as the tactical situation permits. Positioning assets forward—
- Improves response time.
- Reduces road movement.
- Allows the AHS elements to take advantage of the security offered by combat units.

2-165. The thick foliage often makes evacuation by ground more difficult than in other types of terrain. Factors such as the threat, limited road network, and reliance on nonmedical personnel for convoy security make evacuation by air the preferred means. By using the ambulance shuttle system, patients can be transferred from forward-operating ground ambulances to either ground or air ambulances operating further to the rear. In situations where evacuation assets are delayed by various factors (weather or terrain), patients are held for longer periods of time at forward locations. This will dictate the need for additional medical supplies. Army Health System support planners must try to anticipate these delays whenever possible.

The increased disease and infection incidences associated with the jungle environment may worsen the patient's condition; therefore, timely evacuation is essential.

2-166. In some remote and densely foliaged jungles, the only means of evacuation may be by litter. Ambulances may not be practical on trails, unimproved muddy roads, or in swamps. As in mountain operations, there are a higher proportion of litter cases than usual. In the jungle even a slightly wounded Soldier may find it impossible to walk through dense undergrowth. At best, litter teams can carry patients only a few hundred meters over rough terrain before needing rest or relief. Litter carries should be kept as short as possible and medical elements pre-positioned and retained forward.

2-167. Some of the difficulties to consider for MEDEVAC in a jungle environment are—
- The range of frequency modulating radio communications in the jungle may be significantly reduced due to the dense undergrowth, heavy rains, and hilly terrain. The radio transmission range can be extended by using additional radio relays and field expedient antennas.
- There may be few suitable LZs. Many LZs will only be large enough to support one or two helicopters at a time.
- Landing zones may have to be created by clearing brush which takes time and man power.
- Hoist operations may be required more frequently in the thick jungle vegetation where LZs are not available.

DESERT OPERATIONS

2-168. Arid regions make up about one-third of the earth's land surface, a higher percentage than any other type of climate. Desert terrain varies considerably from place to place, with the primary similarity being the lack of precipitation and its consequential effect on vegetation and animal life as well as the terrain. Desert terrain can have mountains, rocky plateaus, sandy dunes, or snow drifts. Rain, when it falls, often causes flooding in low-lying areas or snow and ice in Polar Regions. Wind can have a devastating effect upon AHS operations by destroying equipment and supplies and causing dust storms or blowing snow. Dust or snow storms make navigation and patient treatment difficult. Brown out or white out conditions increase risk to air ambulance operations. Since deserts vary considerably in the type of terrain and temperature, refer to the appropriate OPLAN/OPORD for information on an operational environment.

2-169. Training for desert operations is not significantly different than training for operations in other areas. One consideration is operating in mountainous desert terrain. As many desert areas are in mountainous terrain and because high temperatures increase density altitude, air ambulance evacuation units should be trained in mountain operations. Further, procedures and techniques for evacuation in mountainous terrain must be trained by all MEDEVAC personnel.

2-170. Helicopter landing sites should be chosen with care. Common mistakes made by units when establishing the LZ are—
- Locating the pad relative to the patient and tents, vehicles, and other obstacles. A common tendency is to locate the helipad downwind of the MTFs so that approaches may be made into the wind towards the facility. This causes either a dangerous over flight of the facility or the rotor-wash and dust cloud to encompass the facility. In mountainous deserts, winds normally channel down the valleys and are more predictable along valley floors. A better site selection for an LZ is with the MTF alongside the approach and takeoff zone. Thus, the landing direction is up or down the valley, depending on the airflow, and the MTF is not over flown.
- Marking of helicopter LZs is done so that the site can be seen from the air, but the markings should not be a hazard in themselves. Units requesting MEDEVAC must be prepared to signal the evacuation aircraft upon its arrival. The requesting unit must signal the aircraft to ensure that designated LZ is used.
- Situating LZs in washes, small confined areas between large obstacles, or on routes where vehicles are operating. When operating at higher altitudes air crews may be power limited and lack the ability to land in confined areas, similarly when operating in blowing dust conditions air crews

may not be able to see and avoid obstacles. Refer to FM 3-21.38 for additional guidance on setting up and marking LZs.

2-171. Desert warfare is usually characterized by extended lines of communication which increase evacuation distance and time. The AHS support units may be located further to the rear in the desert or on separated base camps. Establishing an ambulance shuttle system or CCPs is useful. The AHS support units require a greater number of vehicles for operating in deserts than in other environments. Air evacuation by fixed and rotary-wing aircrafts is the preferred method due to their speed and range. Further, using aircraft reduces the load on ground vehicles. Augmentation from higher echelon AHS may also be required to meet the extended evacuation needs.

2-172. The desert environment is challenging and provides little or no protection from enemy air defenses except in the mountainous terrain. Aircraft may have to be flown in such a manner as to reduce its signature in order to reduce its risk to enemy air defense.

OPERATIONS IN COLD REGIONS

2-173. Operations in the extreme cold have many of the limiting factors found in desert operations. The tundra and glacial areas are harsh, arid, and barren. Temperatures may reach lows of 25° Fahrenheit (F) to -40°F (-20° Celsius (C) to -32°C) and when combined with gale-force winds, makes exposure unsurvivable. Refer to ATP 3-90.97 for an in-depth review of cold region operations.

2-174. The greatest environmental detriment to operations is blowing snow. This results in a loss of depth perception from total white conditions. Blowing snow is caused by the wind or by the rotor wash of helicopters; its effect can reduce visibility to zero. Other environmental considerations are as extreme but easier to circumvent. Solid footing is suspect in both dead of winter and in the summer. Snow and ice cover crevasses, holes, and otherwise unstable ground. During the summer, ground transportation is more restricted than in any other environment due to the marsh and muskeg of the arctic tundra. Patients must be sustained for a longer duration due to terrain delays and the lack of direct lines of evacuation.

2-175. Factors to consider for conducting MEDEVAC in arctic operations include the following:
- Arctic warfare is usually characterized by extended lines of communication that increase evacuations distance and time. Establishing an ambulance shuttle system of CCPs and AXPs is useful. Augmentation from higher roles of AHS support may also be required to meet the extended evacuation needs.
- Patient evacuation may have to be sustained for longer periods due to terrain delays and the lack of direct routes of evacuation. During MEDEVAC to an MTF or to an AXP, patients need to be kept as warm as possible, the use of sleeping bags, blankets, or other hypothermia prevention devices is recommended. Warming shelters should also be established along extended lines of evacuation to provide patients and evacuators a means of warming themselves. This allows patients to be monitored for signs of a deteriorating medical condition and provides the personnel performing the evacuation with some relief. Patients with hypothermia require timely evacuation and monitoring throughout the evacuation process.
- The proper storage of medical supplies is essential to prevent loss from freezing or causing further harm to patients. Additional supplies of water should be carried by ambulances and maintained at CCPs, if possible.
- Landing zones must be chosen with care in both winter and summer. During the winter blowing snow takeoffs and landings may require larger LZs then normally used. It can be difficult for litter teams to move patients through snow and over snow drifts; care must be taken not to cause the patient further harm. In the summer movement of patients across tundra and muskeg can make loading air and ground ambulances difficult.

CHEMICAL, BIOLOGICAL, RADIOLOGICAL, AND NUCLEAR CONTAMINATED ENVIRONMENTS

2-176. Medical evacuation and treatment are conducted continuously throughout operations in a CBRN contaminated environment. The AHS commander should have a comprehensive plan which is rehearsed on a periodic basis to ensure the timely evacuation and treatment of CBRN casualties. The number of casualties and their medical conditions, type of contaminant, the size of the land area contaminated, the expected duration of operation, risk assessment and acceptable level of risk, and the number of AHS assets (medical personnel, medical units, and evacuation vehicles and aircraft) initially contaminated will determine the quantity and type of uncontaminated AHS resources, if any, which will be introduced into the contaminated environment to ensure timely medical treatment and evacuation occur. Refer to ATP 4-02.7/MCRP 4-11.1F/NTTP 4-02.7/AFTTP 3-42.3 for additional information on AHS operation in a CBRN environment.

2-177. The commander must take into consideration the number of assets he is willing to commit during evacuation operations in a CBRN-contaminated environment. Since the combinations of evacuation methods are nearly endless, the commander has greater flexibility in tailoring an evacuation plan to meet his particular tactical situation in a CBRN-contaminated environment.

2-178. There are three basic modes of evacuation of patients on the battlefield: personnel, ground vehicles, and aircraft. The following are important considerations for evacuating patients from contaminated environments:

- In using personnel to physically carry the patient, the commander must realize the inherent physical stress involved. Cumbersome mission-oriented protective posture gear needed in a contaminated environment (added to climate, increased workloads, and fatigue) greatly reduces the effectiveness of unit personnel.
- If the commander must send evacuation personnel into a radiological contaminated area, he must establish operational exposure guidance for the MEDEVAC operation. Radiation exposure records are maintained by the unit CBRN noncommissioned officer and are made available to the commander, staff, and surgeon. Based on operational exposure guidance, the commander decides which MEDEVAC assets to send into the contaminated environment.
- Operating MEDEVAC platforms while wearing mission-oriented protective posture gear reduces visibility and hearing which degrades situational awareness and can cause injury or accident. Commanders should incorporate CBRN scenarios to familiarize personnel in these limitations and techniques to mitigate the reduction in effectiveness such gear imposes.

2-179. Commanders should make every effort to limit the number of contaminated evacuation assets while still maintaining a timely and effective medical treatment and evacuation operation. Factors to consider for reducing the impact contaminates have on evacuation assets include the following:

- It is expected that a certain number of both ground and air ambulances will become contaminated. The commander can, therefore, segregate these. This results in the smallest impact on his available assets and the greatest possibility for continuing the patient evacuation mission. Optimize the use of resources, medical or nonmedical, which are already contaminated before employing uncontaminated resources.
- Once a vehicle or aircraft has entered a contaminated area, it is highly unlikely that it will be able to be spared long enough to undergo a complete decontamination. This depends upon the contaminant, the tempo of the battle, and the resources available. Normally, contaminated vehicles (air and ground) have restricted use and are confined to a contaminated environment.
- Introducing uncontaminated aircraft into a contaminated area should be avoided, whenever possible. Ground ambulances should be used instead of air ambulances as long as their use does not adversely affect the patients' medical condition. Ground ambulances are more plentiful and easier to decontaminate. This does not, however, preclude using aircraft in a contaminated environment or in the evacuation of contaminated patients.
- The relative positions of the contaminated area, location of patients, and threat air defense systems determine if and where air assets are to be used. Aviation and medical commanders may choose to restrict one or more air ambulances to the contaminated areas and use ground vehicles to cross the line separating contaminated and clean areas. The ground ambulance can proceed to the receiving MTF with a patient decontamination station. The patient can then be transferred to a clean ground or air ambulance if further evacuation is required. The routes used by ground vehicles to cross between contaminated and clean areas are considered contaminated routes and

should not be crossed by clean vehicles. The effects of wind and time upon contaminants must also be considered.
- The rotor wash of helicopters must always be kept in mind when evacuating contaminated casualties. The intense winds disturb the contaminants in the area and further aggravate the condition by additionally spreading the contaminants. A helicopter must not land too close to a decontamination station (especially upwind) because any trace of contaminants in the rotor wash will compromise the decontamination procedure.
- Evacuation of patients must continue even in a contaminated environment. The commander needs to recognize resource constraints and plan and train to overcome deficiencies.
- Immediate decontamination of aircraft and ground vehicles should be accomplished to minimize crew exposure. Refer to the appropriate equipment technical manual or ATP 3-11.32/ MCWP10-10E.8/NTTP 3-11.37 for equipment decontamination procedures.

BATTLEFIELD OBSCURATION

2-180. Throughout the battlefield, forces acquire and engage targets on visual, laser, and microwave technologies. Friendly and enemy units use obscurants across the battlefield as a combat multiplier. The use of obscurants to mask the combat operations is dictated by the tactical commander. He normally provides the operational guidance for units or elements operating in an area requiring obscuration. Permission to employ obscurants solely to mask MEDEVAC operation may not be approved. However, if the tactical commander's plan indicates that battlefield obscurations are to be employed in a given AO, the medical planner should consider both the advantages and disadvantages posed by their employment. Refer to ATP 3-11.50 for more information on the use of obscurants on the battlefield.

2-181. The medical planner should consider the factors which might impact the use of obscurants in MEDEVAC operations. Factors to consider are the—
- Phase of the tactical operation in which obscurants will be employed.
- Effect on ground and air evacuation routes when operating in an obscured environment (such as limited hours of use, checkpoint or convoy requirements, or the elimination of nap-of-the-earth approaches).
- Potential for exploiting the use of the concealment provided for clearing the battlefield of casualties, especially in defensive operations.
- Potential requirements for obscuration to perform the MEDEVAC mission which would not detract from the tactical capability and requirements.

2-182. The benefit to AHS forces is derived through the tactical commander's use of obscurants to hide friendly tactical maneuvers. This obscuration—
- Prohibits the enemy from knowing how many casualties have been inflicted.
- Aids the movement of medical units and equipment.
- Enhances the ability to resupply forward deployed AHS elements.
- Aids in the tactical deception plan.

GENEVA CONVENTIONS AND THE USE OF OBSCURANTS IN MEDICAL EVACUATION OPERATIONS

2-183. The 1949 Geneva Convention for the Amelioration of the Condition of the Wounded and Sick in Armed Forces in the Field (GWS) provides protection of medical personnel and units from intentional attack so long as they carry out no duties harmful to the enemy (Article 21, GWS). In order to facilitate their identification so as to prevent their intentional attack, medical units, equipment, and personnel are authorized to display the distinctive emblem of the Red Cross (Article 41 and Article 42 GWS). Under tactical conditions, when requirements for concealment outweigh those for recognition, all distinctive emblems may be obscured or removed from medical equipment if ordered by a brigade or higher commander and authorized by AR 750-1. Display of the distinctive emblem is not required to afford the right against intentional attack, attack of medical units, equipment, and personnel not displaying the distinctive emblem are prohibited if opposing forces realize that the forces about to be attacked are medical units performing humanitarian duties.

2-184. The use of obscurants in MEDEVAC operations does not differ from the use of camouflage techniques and is not prohibited by the GWS. Its only effect will be to obscure the identity of units as they perform their humanitarian mission. Given the lethality of the modern battlefield, however, it would be difficult, if not impossible, to say that such obscuration of these units, equipment, and personnel would necessarily increase their risk from unintentional attack.

2-185. It is recognized that, with the advent of precision-guided munitions and electro-optical or laser target acquisition devices, there will be a substantial use of obscurants on the battlefield as a result of normal combat operations. The legitimate use of obscurants by combatants to thwart the accuracy of precision guided munitions may increase the risk to units and equipment not employing obscurants. This may possibly place medical units and equipment at greater risk if they fail to employ them. Further, MEDEVAC operations will have to be carried out on the battlefield as medical personnel find it, which will include obscurants employed for combat operations.

USE OF SMOKE IN AIR AMBULANCE AND HOIST RESCUE OPERATIONS

2-186. Smoke and obscurants have several uses during air ambulance and hoist rescue operations. Some of these uses can be very beneficial to the pilot and crew in locating the patient, communicating with the ground personnel, and determining environmental conditions. Colored smoke grenades (M18 smoke grenade) can be used effectively in air ambulance and overland hoist rescue operations to—

- Identify the landing site. Colored smoke is an excellent daytime marking method. The smoke generated from a smoke grenade is difficult to detect at more than 2 to 3 miles away, but an aircraft in the area should have little difficulty in noting its location. As more than one unit may be operating in a given area, it is important that the unit requesting a MEDEVAC mission be able to signal the aircraft as to the correct landing site to use. When a unit employs colored smoke to mark a landing site, the aircrew should identify the color and confirm it with ground personnel.
- Reduce electronic signature. Radio communications produce an electronic signature. The electronic signature created from a prolonged transmission to guide an air ambulance to the landing site may not be an acceptable tactical risk. The transmission time required to signal the aircraft using smoke is limited, thereby reducing the electronic signature.
- Determine surface wind direction. The employment of smoke at the landing site also enables the aircrew to determine the wind direction.
- Provide concealment. In some environmental conditions (such as desert operations), the phenomenon of inversion occurs. When this occurs, obscurants used in normal combat operations may provide an upper layer of smoke under which the air ambulance can operate.

2-187. The use of obscurants on air ambulance operations can be a disadvantage if incorrectly employed or generated. Obscurants can hide the landing site and make nap-of-the-earth approaches unfeasible. Further, battlefield obscurations can force aircraft to fly at higher than planned heights. This may increase the risk of being acquired by the enemy.

2-188. In overwater hoist rescue operations, the employment of smoke, from a marine smoke and illumination signal device, can be used for marking the patient pickup area, for determining surface wind conditions and for aiding in spatial orientation. The smoke employed by the aircrew must not interfere with the conduct of the operation or mask the location of the individual to be rescued.

EMPLOYMENT OF OBSCURANTS IN GROUND MEDICAL EVACUATION OPERATIONS

2-189. The employment of obscurants during ground evacuation operations must be in consonance with the tactical commander's plan. Obscurants can mask MEDEVAC operations on the battlefield, but must not interfere with the tactical mission. In all combat operations, but especially in urban operation, obscurants can be employed to conceal—

- Movement across open areas.
- Extraction of casualties from vehicles or buildings.
- Entry and exit into or out of structures.

This page intentionally left blank.

Chapter 3
Medical Evacuation Resources

This chapter discusses the mission, functions, and capabilities of MEDEVAC units and elements, as specified in the unit's table of organization and equipment. It also discusses the mission command headquarters to which they are assigned. The discussion of each MEDEVAC organization also includes their limitations and dependencies.

SECTION I — MANEUVER BATTALION MEDICAL PLATOON AMBULANCE SQUAD

3-1. The maneuver battalion medical platoon ambulance squad enhances the battalion's ability to sustain initiative by sustaining medical proximity and mobility with the maneuver force. The maneuver battalion is unencumbered by casualties due to timely and efficient MEDEVAC.

3-2. The medical platoon is part of the BCT's maneuver battalion's headquarters and headquarters company. The medical platoon consists of a platoon headquarters section, a medical treatment squad, combat medic section, and an ambulance squad. The maneuver battalion medical platoons are discussed in detail in ATP 4-02.3.

MEDICAL TREATMENT PLATOON AMBULANCE SQUADS

3-3. The primary mission of the maneuver battalion's medical platoon ambulance squad is to provide ground ambulance evacuation support from supported infantry/armored companies or from POI back to a CCP or to the Role 1 MTF/BAS. They also provide area support to other elements (which do not have organic MEDEVAC resources) operating in their AO.

AMBULANCE OR EVACUATION SQUAD

3-4. The ambulance squad of the medical platoon is organized into ambulance teams. The number of ambulance teams in the ambulance squad varies and is based on the type of parent organization. The infantry and infantry (airborne) battalion's ambulance squads have high mobility multipurpose wheeled vehicle (HMMWV) ambulances. The armored combat maneuver battalion's medical platoon ambulance squad has legacy M113 track ambulances or Armored Multi-Purpose Vehicle, Medical Evacuation. The Stryker BCT infantry battalions' medical treatment platoon's evacuation squad has two teams or four Stryker wheeled armored ambulances, referred to as the Stryker MEDEVAC vehicle.

AMBULANCE TEAM

3-5. The ground ambulance team is the basic MEDEVAC component used within the BCTs and within the EAB AOs. An ambulance team consists of two ambulances, each ambulance has three medical personnel (an emergency care sergeant or health care specialist and two ambulance/aide drivers). Each of these individual ambulances is referred to as an element. These ground ambulance teams provide MEDEVAC from the POI to supporting MTFs while ensuring the continuity of care en route.

3-6. The primary function of the ambulance team is to collect and treat the sick, injured, and wounded Soldiers on the battlefield and to provide MEDEVAC support from the POI, CCP, or AXP to the supporting MTF.

3-7. Ambulance team members are also responsible for performing operator maintenance on assigned ambulances. Specific duties of the ambulance team are to—
- Find and collect the wounded.
- Administer tactical combat casualty care as required.
- Initiate the DD Form 1380 as required.
- Provide patient care en route.
- Prepare patients for air and ground MEDEVAC.
- Provide MEDEVAC of wounded or injured Soldiers from the POI to supporting MTF or AXP.
- Provide emergency movement of medical personnel and emergency delivery of blood, medical supplies, and medical equipment.
- Assist in the care and transport of combat and operational stress reaction casualties.
- Resupply company and platoon combat medic, when required.
- Serve as messengers within medical channels.
- Operate the vehicle and maintain contact with supported and supporting elements.
- Maintain operational readiness of the ambulance.

SECTION II — EVACUATION PLATOON—MEDICAL COMPANY (BRIGADE SUPPORT BATTALION) AND AMBULANCE PLATOON—MEDICAL COMPANY (AREA SUPPORT)

3-8. Medical companies provide Role 2 MTF and MEDEVAC. They are assigned to the BCTs as BSMC and to the MMBs as MCAS.

MEDICAL COMPANIES

3-9. The BSMC provides direct support medical care to supported maneuver battalions with organic medical platoons. The BSMC also provides Roles 1 and 2 medical treatment on an area basis to those units without organic medical assets operating in the brigade support area. The MCAS operate Role 2 MTFs and provide MEDEVAC support for units at EAB. In addition, medical companies provide Role 1 area medical support to units without organic medical personnel and may augment/reinforce maneuver medical platoons/sections with treatment and MEDEVAC support.

MEDICAL COMPANY (BRIGADE SUPPORT)

3-10. The evacuation platoons assigned to the BSMC provide ground ambulance evacuation support from the supported BCT or from the POI to the supporting MTF.

3-11. The BSMC is designed to support the BCT. The basis of allocation for the BSMC is one per brigade support battalion. The mission of the BSMC is to provide AHS support to all BCT units operating within the brigade AO. The BSMC locates and establishes its company headquarters in the brigade support area and operates a Role 2 MTF and may operate Role 1 MTFs on an area support basis for units that do not have organic medical assets. The company provides mission command for its organic and attached or OPCON medical augmentation elements. The BSMC may be augmented with an FST or FRST providing the company limited surgical capability. See ATP 4-02.5 for additional information on the FST. The BSMC is organized into a company headquarters, the brigade medical supply office, a preventive medicine section, a mental health section, an evacuation platoon made up of a forward and area evacuation squad, and a medical treatment platoon made up of a platoon headquarters, medical treatment squad, area support squad, medical treatment squad (area), and a patient holding squad.

3-12. The evacuation platoon performs ground MEDEVAC and en route patient care for the supported units. The evacuation platoon consists of a platoon headquarters, a general support evacuation squad (area), and a direct support evacuation squad (forward). The platoon employs five evacuation teams (or ten ambulances), with three teams in the evacuation squad (forward) and two teams in the evacuation squad (area). A complete discussion on the organization, mission, and functions of the BSMC is provided in ATP 4-02.3. In the Stryker

BCT and Armor BCT ground ambulances assigned to the evacuation squad (forward) must be able to match the maneuverability, defendability, and survivability of the supporting maneuver force. In the infantry brigade combat team, the BSMC shares the same evacuation vehicle and level of protection as their supported maneuver battalions.

MEDICAL COMPANY (AREA SUPPORT)

3-13. The MCAS performs functions similar to those of the BSMC. The basis of allocation for the MCAS is one per 10,000 non-BCT Soldiers supported in the committed brigade, division headquarters, corps headquarters and committed theater Army. The MCAS is employed primarily in support of units at EAB. They are deployed to a geographical area to provide area HSS, or may be deployed to provide HSS for designated units. The MCAS also establishes its Role 2 MTF in a secure location centrally located for supported units. Medical treatment squads/teams of the MCAS may be deployed to establish Role 1 MTF and provide HSS support to concentrations of nondivisional units that do not have organic medical capabilities. The MCAS is organized into a company headquarters, a mental health section, an ambulance platoon, and a medical treatment platoon made up of a platoon headquarters, medical treatment squad, area support squad, area support medical treatment squad, and a patient holding squad.

3-14. The ambulance platoon performs ground MEDEVAC and en route patient care for supported units. The ambulance platoon consists of a platoon headquarters, four ambulance squads (or eight ambulance teams), one HMMWV control vehicle, and eight HMMWV ambulances. A complete discussion on the organization, mission, and functions of the MCAS is provided in ATP 4-02.3.

ORGANIZATIONS

3-15. The BSMC normally establishes its Role 2 MTF in the brigade support area; whereas, the MCAS normally establishes its Role 2 MTF in a division support area, corps support area, and at theater level. The medical company provides direct and area support ground ambulance evacuation for supported units. It coordinates/requests air ambulance support for air evacuation of patients from supported Role 1 MTFs/BASs, POI, CCP, and AXPs. The organizational structure, number and type of ambulances assigned to an ambulance/evacuation platoon may differ depending on their location on the battlefield and the types of units supported. For example, a BSMC in support of an Armored BCT has a mix of armored and wheeled evacuation vehicles while a BSMC in support of the Stryker BCT has only wheeled evacuation vehicles.

AMBULANCE/EVACUATION PLATOON

3-16. The ambulance/evacuation platoon provides mission command for ambulance platoon operations. The ambulance platoon headquarters element maintains communications to direct ground ambulance evacuation of patients. It provides ground ambulance evacuation support for supported maneuver battalions and for supported units operating in the consolidation or support areas. The ambulance headquarters element performs route reconnaissance, develops and issues graphic overlays to all its ambulance teams. It also coordinates and establishes AXPs for both air and ground ambulances, as required.

AMBULANCE SQUAD

3-17. Ambulance squads provide ground ambulance evacuation of patients from supported BASs/unit aid stations back to the Role 2 MTF that is located in the brigade support area. An ambulance squad consists of two ambulance teams (two ambulances, wheeled or tracked vehicles). Ambulance squad personnel—
- Perform tactical combat casualty care, evacuate patients, and provide for their continued care en route.
- Operate and maintain assigned communication and navigational equipment.
- Perform preventive maintenance checks and services on ambulances and associated equipment.
- Maintain supply levels for the ambulance medical equipment sets.
- Ensure that appropriate property exchange of medical items (such as litters and blankets) is made at sending and receiving MTFs.

Chapter 3

- Prepare patients for air and ground MEDEVAC.
- Provide MEDEVAC of wounded or injured Soldiers from the POI to supporting MTF, CCP or AXP.
- Initiate the DD Form 1380 as required.
- Provide emergency movement of medical personnel and emergency delivery of blood, medical supplies, and medical equipment.
- Assist in the care of combat and operational stress reaction casualties.
- Maintain operational readiness of the ambulance.
- Resupply company and platoon combat medics, when required.
- Operate the vehicle and maintain contact with supported and supporting elements.
- Find and collect the wounded.
- Serve as messengers within medical channels.

SECTION III — MEDICAL COMPANY (GROUND AMBULANCE)

3-18. The medical company, ground ambulance has the ability to increase the medical evacuation capacity in a supported area without the additional Role 2 treatment assets that are part of the medical company, area support.

3-19. The AHS is a continuum of increasing roles of care extending from the POI through the CONUS base. All sick, injured, and wounded patients must be evacuated from the battlefield in the shortest possible time to the MTFs that can provide the required treatment. The medical company (ground ambulance) (see Figure 3-1) serves as one of the primary means of evacuating patients from the battlefield by ground.

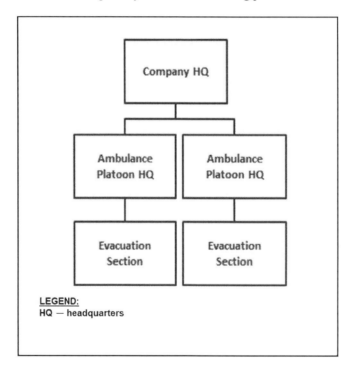

Figure 3-1. Medical company, ground ambulance

MISSION

3-20. The mission of the medical company (ground ambulance) is to provide MEDEVAC within the theater of operations.

ASSIGNMENT

3-21. The medical company (ground ambulance) is normally assigned or attached to the MMB or a MEDBDE (SPT) for mission command. In-depth information on these commands can be found in FM 4-02.

EMPLOYMENT

3-22. The medical company (ground ambulance) is employed in EAB to provide area support. It is tactically located where it can best control its assets and execute its patient evacuation mission.

CAPABILITIES

3-23. At Role 1, the unit is capable of providing—
- A single-lift capability for the company is 96 litter patients or 192 ambulatory patients.
- Medical evacuation from BCT medical companies and MCASs to supporting hospitals.
- Reinforcement of BCT medical company evacuation assets.
- Reinforcement of covering forces and deep battle operations.
- Movement of patients between hospitals and aeromedical staging facilities, aeromedical staging squadrons, railheads, or seaports in the EAB.
- Area evacuation support beyond the capabilities of the MCAS.
- Emergency movement of medical personnel and supplies.
- Medical evacuation of wounded or injured Soldiers from the POI to supporting MTF.

DEPENDENCY

3-24. This unit is dependent upon the appropriate elements of the theater Army for—
- Religious, financial management, legal, personnel, mortuary affairs, and administrative services.
- Viable communications systems for mission command and adequate road networks.
- Laundry, shower, and clothing repair.
- Generator, communication equipment, and communication security equipment maintenance.
- AHS support to include hospitalization.

ORGANIZATION AND FUNCTIONS

3-25. This company is organized into a company headquarters section, two ambulance platoons each with a headquarters, and evacuation sections (see Figure 3-1).

3-26. The company headquarters provides mission command, communications, administration, field feeding, and logistical support (to include maintenance) for the subordinate ambulance platoons.

3-27. The ambulance platoon headquarters provides mission command for the subordinate ambulance squads.

3-28. The evacuation section operates ambulances and provides en route medical care for patients. Each evacuation section consists of 12 ambulances, each with a two-man crew. The members of the squad maintain the level of expendable Class VIII supplies in the ambulance medical equipment set by reconstituting supplies from medical companies or hospitals when they pick up or drop off patients. They are also responsible for performing operator maintenance on assigned vehicles.

Chapter 3

SECTION IV — MEDICAL COMPANY (AIR AMBULANCE) 15 HH-60

3-29. The medical company (air ambulance) 15 HH-60 aircraft (see figure 3-2) provides MEDEVAC for all categories of patients consistent with evacuation precedence and other operational considerations within the division.

3-30. The medical company (air ambulance) falls under the GSAB which will provide aircraft maintenance and logistics support, aviation communications, and real-time operational picture associated with today's combat environment. Additional information on the GSAB can be found in FM 3-04 and ATP 3-04.1.

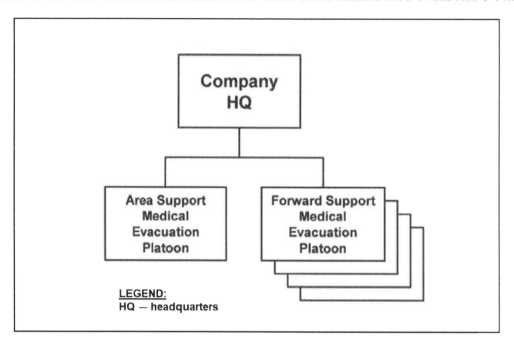

Figure 3-2. Medical company (air ambulance) 15 HH-60

MISSION

3-31. The mission of the medical company (air ambulance) (15 HH-60s) is to provide aeromedical evacuation support within the brigade and corps.

ASSIGNMENT

3-32. The medical company (air ambulance) (15 HH-60s) is organic to the GSAB for mission command.

EMPLOYMENT

3-33. The medical company (air ambulance) (15 HH-60s) is employed as needed in the theater, corps, division, or EAB. It is tactically located where it can best control its assets and executes its patient evacuation mission.

CAPABILITIES

3-34. The medical company (air ambulance) (15 HH-60s) provides—
- Fifteen helicopter ambulances to evacuate critically wounded or other patients consistent with evacuation priorities and operational considerations, while providing en route care from points as

far forward as possible, to the appropriate MTF. Total lift capability utilizing all assigned aircraft is 90 litter patients or 105 ambulatory patients, or some combination thereof.
- One area support MEDEVAC platoon (three aircraft) that will normally locate with the company headquarters. Four FSMP (three aircraft each) that can be independently or group deployed.
- Air crash rescue support.
- Expeditious delivery of whole blood, biological, and medical supplies to meet critical requirements.
- Rapid movement of medical personnel and accompanying equipment/supplies to meet the requirement for mass casualty situations, reinforcement/reconstitution, or emergency situations.
- Movement of patients between hospitals, aeromedical staging facilities, hospitals ships, casualty receiving and treatment ships, seaports, and railheads in the brigade AO.
- Military working dog evacuation.
- Support combat search and rescue.

DEPENDENCY

3-35. The medical company (air ambulance) (15 HH-60s) is dependent on—
- Appropriate elements within the theater for AHS support.
- The headquarters and headquarters company of the GSAB for religious, legal, finance, personnel and administrative services, and unit CBRN support.
- The forward support company of the GSAB for Class III, automotive and generator maintenance, and field feeding.
- The aviation support company of the GSAB for aviation unit maintenance of organic aircraft, including unit-level supply support for aircraft (Class IX).
- United States Air Force weather team in the headquarters and headquarters company of the CAB for air weather service support.

ORGANIZATION AND FUNCTIONS

3-36. The company headquarters provides mission command of all area support and forward support MEDEVAC operations, and provides logistical and administrative support for the company.

3-37. The area support MEDEVAC platoon provides area support medical evacuation support within the brigade AO.

3-38. The four forward support MEDEVAC platoons provide a task-organized means for air ambulance support of brigades. They also provide emergency movement of medical personnel and emergency delivery of whole blood, biological, and medical supplies and equipment.

SECTION V — MEDICAL COMPANY (AIR AMBULANCE) UH-72 LIGHT UTILITY HELICOPTER

3-39. The medical company (air ambulance) UH-72 aircraft (see Figure 3-3 on page 3-8) provides MEDEVAC for all categories of patients consistent with MEDEVAC precedence.

3-40. The medical company (air ambulance) UH-72 falls under the aviation security and support battalion which will provide aircraft maintenance and logistics support, aviation communications, and real-time operational picture associated with its medical support mission.

Chapter 3

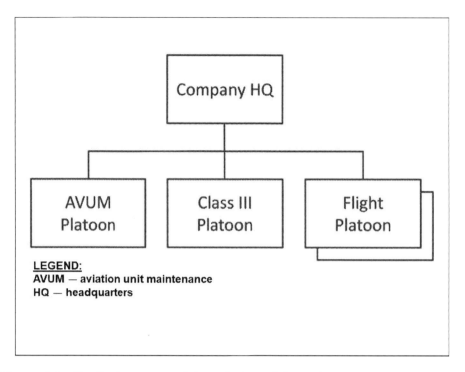

Figure 3-3. Medical company (air ambulance) (UH-72 light utility helicopter)

MISSION

3-41. The mission of the medical company (air ambulance) (light utility helicopter) is to provide aero-medical evacuation in support of homeland security/defense requirements to include support of selected overseas and outside of the CONUS operations.

ASSIGNMENT

3-42. The medical company (air ambulance) (light utility helicopter) is organic to the aviation security and support battalion for mission command.

EMPLOYMENT

3-43. The medical company (air ambulance) (light utility helicopter) provides geographically dispersed, readily available, light utility air ambulance capability for Federal or State authorities and CCDRs. Majority of operations will occur in the U.S. and its territories in support of homeland security/defense. The company can also deploy and operate worldwide in permissive environments.

CAPABILITIES

3-44. The medical company (air ambulance) (light utility helicopter) provides—
- Eight helicopter ambulances to evacuate critically wounded or other patients consistent with evacuation priorities and operational considerations while providing en route care, from points as far forward as possible, to the appropriate MTF. Total lift capability utilizing assigned aircraft is 16 litter patients or 40 ambulatory patients or a combination of eight litter and 24 ambulatory patients.
- Two MEDEVAC flight platoons with four helicopters in each, which can be independently deployed.

- Expeditious delivery of whole blood, biological, and medical supplies to meet critical requirements.
- Rapid movement of medical personnel, and accompanying equipment/supplies, to meet the requirements for mass casualty, reinforcement/reconstitution, or emergency situations.
- Movement of patients between aeromedical staging facilities.
- Immediate and deliberate search and rescue operations.
- Air ambulance support to U.S. Northern Command area of responsibility and as required, to State Adjutant Generals.
- Air ambulance support to respective Weapons of Mass Destruction Civil Support Teams during CBRN incidents.
- Continuous day and night aviation operations during visual and marginal weather conditions and limited aviation operations during instrument weather conditions.
- Unit-level logistical support to include supply support, field-level maintenance (ground), communication maintenance (except communication security equipment), and aircraft Class III/IX.
- Aviation unit maintenance for eight organic UH-72 MEDEVAC aircraft.

DEPENDENCY

3-45. The medical company (air ambulance) (light utility helicopter) is dependent on—
- Appropriate elements within the theater for religious, legal, AHS support, finance, and personnel and administrative services.
- Headquarters and headquarters company, aviation security and support battalion, for unit-level logistical support.
- United States Air Force weather team in the headquarters and headquarters company of the CAB for air weather services support.

ORGANIZATION AND FUNCTIONS

3-46. The medical company (air ambulance) (light utility helicopter) is organized into a company headquarters section, a Class III platoon, two MEDEVAC flight platoons, and an aviation unit maintenance platoon.

3-47. Company headquarters section provides mission command, coordinates activities of the security and support aviation company, unit-level administrative, and CBRN support. The company headquarters is comprised of the flight operation section, supply section, automotive maintenance section, and communications section.

3-48. The Class III platoon plans, coordinates, and ensures responsive execution of Class III support for company aircraft. It is comprised of the platoon headquarters and Class III section, which dispenses fuel into the aircraft and performs operator maintenance on all vehicles and refueling equipment.

3-49. The two MEDEVAC flight platoons provide eight light utility helicopter MEDEVAC aircraft, associated flight crews, and personnel for accomplishment of the company's air MEDEVAC mission.

3-50. Aviation unit maintenance platoon headquarters provides mission command activities for the aviation unit maintenance platoon's quality assurance section, production control section, aircraft maintenance section, aircraft component repair section, and avionic/electrical section. These sections perform all associated aircraft inspections, maintenance, and repair activities.

This page intentionally left blank.

Chapter 4
Theater Evacuation Policy

This chapter discusses the factors and impacts of a theater evacuation policy.

ESTABLISHING THE THEATER EVACUATION POLICY

4-1. The theater evacuation policy is established by the Secretary of Defense, with the advice of the Joint Chiefs of Staff and upon the recommendation of the theater commander. The medical commander may recommend changes to the theater evacuation policy to adjust the patient flow in order to surge capability when necessary. The policy establishes the length in days of the maximum period of noneffectiveness (hospitalization and convalescence) that patients may be held within the theater for treatment. This policy does not mean that a patient is held in the theater for the entire period of noneffectiveness. A patient who is not expected to be ready to return to duty within the number of days established in the theater evacuation policy is evacuated to the CONUS or other safe haven. This is done, providing that the treating physician determines that such evacuation will not aggravate the patient's disabilities or medical condition. For example, a theater evacuation policy of seven days does not mean that a patient is held in the theater for six days and then evacuated. Instead, it means that a patient is evacuated as soon as possible after the determination is made that the Soldier cannot be returned to duty within seven days following admission to a Role 3 hospital.

TEMPORARY REDUCTIONS

4-2. When unplanned increases in patients occur (due perhaps to an epidemic or heavy combat casualties), a temporary reduction in the policy may be necessary. This reduction is used to adjust the volume of patients being held in the theater hospital system. A reduction in the evacuation policy increases the number of patients requiring intertheater evacuation and the demand for evacuation assets. This action is necessary to relieve the congestion caused by the patient increases. A decrease in the theater evacuation policy increases evacuation asset requirements.

ESSENTIAL CARE IN THEATER AND LENGTH OF POLICY

4-3. The hospitalization capability in the AO offers essential care to either return the patient to duty (within the theater evacuation policy) and/or stabilize the patient so he can tolerate evacuation to a definitive care facility outside the AO. As a result of a reduced medical footprint in a deployed setting, medical care delivery to our forces is provided under the concept of essential care in theater. This concept provides for essential and stabilizing care being rendered within the theater, with the patient being evacuated to CONUS or another safe haven for definitive, rehabilitative, and convalescent care.

4-4. The time period established by the theater evacuation policy starts on the date the patient is admitted to the first Role 3 hospital. The total time a patient is hospitalized in the theater (including transit time between MTFs) for a single, uninterrupted episode of illness or injury should not exceed the number of days stated in the theater evacuation policy. Though guided by the evacuation policy, the actual selection of a patient for evacuation is based on clinical judgment as to the patient's ability to tolerate and survive the movement to the next role of care.

EXCEPTION TO POLICY

4-5. An exception to the theater evacuation policy may be required with respect to low density military occupational specialty skills such as SOF personnel within the theater. Retaining these personnel within the theater for an extended period of time is possible if the medical resources are available within the theater to treat their injuries and provide for convalescence and rehabilitation. If retention within the theater would result in a deterioration of their medical condition or would adversely impact their prognosis for full recovery, they are evacuated from the theater for definitive care. In addition, a nontransportable patient may require an extended stay in order to stabilize him or her enough to be evacuated from theater. A *nontransportable patient* **is a patient whose medical condition is such that he could not survive further evacuation to the rear without surgical intervention to stabilize his medical condition.**

FACTORS DETERMINING THE THEATER EVACUATION POLICY

4-6. To fully understand how the theater evacuation policy affects AHS operations, the medical operator should be aware of the factors that influence the establishment of this policy.

EVACUATION POLICY CONSIDERATIONS

4-7. When determining the length of the theater evacuation policy, several considerations should be taken into the planning process. Paragraphs 4-8 through 4-20 discuss the various considerations of the theater evacuation policy.

Medical Providers

4-8. To medical providers, such as physicians and dentists who are engaged directly in patient treatment and decisions relating to patient disposition, the evacuation policy means that there is a maximum period which clinical staffs may complete the necessary treatment needed to return the patient to full duty within the theater. If the theater policy is seven days and a full return to duty can be predicted within that time, the patient is retained in the theater. If the patient cannot be returned to full duty within seven days, the patient is evacuated out-of-theater as early as clinically prudent.

Medical Planners and Staff

4-9. Medical staffs at all levels are required to maintain situational understanding of the capabilities and status of the medical assets that are available to them either in a direct or area support role. The medical staff compiles and analyzes hospital bed status among other factors to determine the number and distribution of hospital beds available and required in the theater. The medical staff also compiles and tracks any special medical capability within the theater. A lack of a specialized care may also prompt the evacuation of a patient.

4-10. The medical planner has a management tool which, when properly adjusted and used, provides casualty estimation based off of adjustable parameters such as length of operation and engagements, weather, terrain, and country of operation. Medical planners synchronize medical treatment with ground and air evacuation assets to sustain proximity and provide flexibility to meet changing requirements.

United States Air Force Planner

4-11. The USAF has dedicated units that coordinate aeromedical evacuation assets that provide both intra- and intertheater patient movements. *Aeromedical evacuation* is the movement of patients under medical supervision to and between medical treatment facilities by air transportation (JP 4-02). Refer to Chapter 6 of this publication and JP 4-02 for more information on United States Transportation Command (USTRANSCOM) and Air Mobility Command assets.

NATURE OF THE OPERATION

A major factor in determining the theater evacuation policy is the nature of the operations. Operations of short duration and with a low potential for conflict may require a short evacuation policy due to limited available medical support. Operations of long duration with significant combat operations could require a longer evacuation policy in order to return as many personnel to duty in theater as possible as opposed to evacuating the patients out of theater. A longer evacuation policy has a reduced demand on evacuation assets and will retain patients in theater longer increasing the potential for their return to their unit. As a result of a longer theater evacuation policy, there is a greater requirement for bed space and medical treatment at

Role 2 and Role 3, which reduces the mobility and capabilities of the MTFs. A shorter theater evacuation policy will increase the demand on evacuation assets but reduce occupancy of hospital beds which increases the mobility of MTFs and provides holding capability in preparation of major operations. When the majority of anticipated patients are from combat related trauma, a shorter evacuation policy may be required in order to quickly move patients out of theater and sustain mobility. This is especially true when the number and capabilities of Role 2 and 3 MTFs would be quickly exhausted.

NUMBER AND TYPE OF PATIENTS

4-12. Another factor affecting the policy is the number and types of patients anticipated and the rate of patients that return to duty. Admission rates vary widely in different geographical areas of the world and in different types of military operations.

EVACUATION MEANS

4-13. The means (quantity and type of transportation) available for evacuation of patients from the theater to CONUS is an essential factor impacting the evacuation policy.

AVAILABILITY OF IN-THEATER RESOURCES

4-14. Limitations of all AHS resources, such as insufficient number and types of medical units in EABs to support the BCTs and an insufficient amount of medical and nonmedical logistics, will have an impact on the theater evacuation policy. A shorter theater evacuation policy will reduce some demand on limited resources such as Class I (subsistence) to sustain patients by reducing the number of patients held in MTFs. The more limitations (or shortages), the shorter the theater evacuation policy.

IMPACT OF THE EVACUATION POLICY ON ARMY HEALTH SYSTEM REQUIREMENTS

4-15. The length of the theater evacuation policy can have multiple impacts on the AHS and military health system. The following paragraphs explain what those effects are on patients, evacuation assets, and MTFs.

SHORTER EVACUATION POLICY

4-16. A shorter theater evacuation policy—
- Results in fewer hospital beds required in the theater and a greater number of beds required elsewhere.
- Creates a greater demand for intertheater USAF and intratheater evacuation resources.
- Increases the requirements for replacements to meet the rapid personnel turnover which could be expected, especially in combat units.

LONGER EVACUATION POLICY

4-17. A longer theater evacuation policy—
- Results in a greater accumulation of patients and a demand for a larger AHS infrastructure in the theater. It decreases bed requirements elsewhere.
- Increases the requirements for medical logistics (medical supplies, equipment, and equipment maintenance) and nonmedical logistics support.
- Increases the requirements for hospitals, engineer support, and all aspects of base development for deployed AHS force. (It demands the establishment of a larger number of hospitals within the theater and may require medical specialty augmentation.)
- Provides for a greater proportion of patients to return to duty within the theater and, thus, reduces the loss of experienced manpower.
- A longer evacuation policy may decrease the demand on the intratheater evacuation assets and system.

Chapter 4

4-18. The concept of essential care in theater does not support longer evacuation policies as the deployed hospitals are not designed to provide definitive, rehabilitative, and convalescent care/services. If the theater evacuation policy is extended in theaters operating under the essential care in theater concept, augmentation of medical specialty resources will be required.

PATIENT STABILIZATION

4-19. The evacuation policy has no impact on the patient stabilization period for movement. This period is known as the evacuation delay. It is the period of time planned for between the time of patient reporting and the time of MEDEVAC of the patient to the next role of care. Evacuation delays normally range from 24 to 72 hours and are designated by the Army theater surgeon.

Chapter 5
Operational and Tactical Evacuation Planning

A comprehensive MEDEVAC plan is essential to ensure effective, efficient, and responsive MEDEVAC is provided to all wounded, injured, and ill Soldiers in the AO. The Army MEDEVAC plan flows from the CCDR's guidance and intent, incorporates all missions and tasks directed by the CCDR to be accomplished, and is synchronized with supporting and supported units. Examples of a MEDEVAC OPLAN and OPORD are contained in Appendix B.

SECTION I — THEATER MEDICAL EVACUATION PLANNING RESPONSIBILITIES

5-1. All MEDEVAC planning within the joint operations area reflects the CCDR's guidance and intent. Planning at the joint level is governed by JP 5-0. Guidance on MEDEVAC in joint operations is addressed in JP 4-02.

JOINT PLANNING

5-2. When directed by the CCDR, Army MEDEVAC assets may be tasked to support other than Army forces engaged in the execution of the joint mission. These additional support missions will be clearly articulated in the CCDR's OPLAN and OPORD. The theater Army surgeon, with the advice of the senior MEDEVAC planner, will coordinate and synchronize these support operations with the combatant command surgeon, joint task force surgeon, and the other Services and/or multinational partners as required to ensure that a comprehensive and effective, efficient, and responsive plan is developed and implemented.

MEDICAL COMMAND AND CONTROL ORGANIZATIONS

5-3. Army evacuation assets support the full range of all Army operations and are capable of supporting Army forces engaged in the execution of the joint mission, as directed. Major combat operations supported by a theater Army will have a deployed MEDCOM (DS). The mission command coordination and orders flow for AHS is illustrated in Figure 5-1 on page 5-2. Evacuation planning should occur at each major level of command in order to effectively execute mission command, and influence over evacuation planning.

Chapter 5

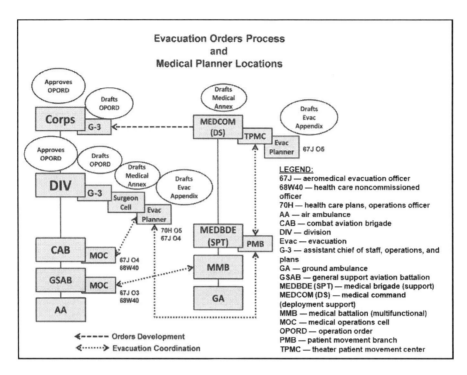

Figure 5-1. Mission command organizations for coordination and orders flow

5-4. Some early entry operations may not require the deployment of a theater or a brigade mission command structure. In these cases, MEDEVAC planning, coordination, and synchronization are initiated by the most senior medical command/staff level (such as the division surgeon, MEDBDE [SPT], GSAB, and/or MMB). The senior leaders responsible for air and ground MEDEVAC provide estimates and recommendations for the plan. This plan is normally provided as part of Annex F (sustainment) in the MEDEVAC subparagraph of the Appendix 3 (health system support).

5-5. Dedicated Army rotary-wing aircraft (air ambulances) are under the mission command of the GSAB within the current Army aviation structure. Ground ambulances remain under the mission command of medical units/elements that are organic to movement and maneuver, fires, and sustainment units. The medical company (ground ambulance) is under the control of the MMB.

5-6. To ensure an integrated MEDEVAC system, synchronization of air with ground ambulance assets is essential. To affect this synchronization, the medical command and control headquarters must maintain influence over the execution of the MEDEVAC plan. This influence is accomplished through the orders process.

THEATER PATIENT MOVEMENT CENTER

5-7. In the design of the MEDCOM (DS), the MRO staff was reevaluated, realigned, and augmented with medical evacuators to more effectively coordinate and synchronize the various ground and air evacuation aspects of the operation and to supervise and facilitate the medical regulating functions. This staff section was renamed the theater patient movement center (TPMC). The addition of medical evacuators enables this section to develop, coordinate, and synchronize the MEDEVAC portion of the MEDCOM (DS), AHS OPLAN/OPORD between MEDCOMs, and aviation mission command structures. Additionally, the TPMC maintains 24-hour coordination and management responsibility for patient regulating and administration within the MEDCOM (DS) AO, and directs patient evacuation between facilities within its AO, as required, avoiding patient overflow. This staff is responsible for preparing patient statistical reports for the command. This section also coordinates with the supporting theater patient movement requirements center (TPMRC) for the transfer of patients out of the AO. The TPMRC regulates the flow of patients out of the AO to an

Operational and Tactical Evacuation Planning

MTF capable of providing the required specialty care and arranges for the transport of these patients by USAF aeromedical evacuation assets.

PATIENT MOVEMENT BRANCH

5-8. As with the MEDCOM (DS), the MRO of the MEDBDE (SPT) was designed in a similar fashion to enhance the capabilities for MEDEVAC planning within the MEDBDE (SPT) AO. This staff section was named the patient movement branch. This office is responsible for coordinating and synchronizing MEDEVAC operations with the MEDCOM (DS) TPMC and the division surgeon cell. It regulates patients within its subordinate MTFs and arranges for transportation to transfer patients between facilities to relieve surgical backlogs, to obtain specialty care, and to ensure bed availability prior to start of operations, as appropriate. It also coordinates with the MEDCOM (DS) TPMC and the TPMRC for the evacuation of patients out of theater. This office maintains a 24-hour coordination and management responsibility for patient regulating and administration within the MEDBDE (SPT) AO. It also prepares patient statistical reports for the MEDBDE (SPT) commander. In the absence of a MEDCOM (DS) in theater, the patient movement branch assumes the duties of the MEDCOM (DS's) TPMC.

SECTION II — PLANNING PROCESS

5-9. Operation plans and orders are the means by which commanders express their visualization, commander's intent, and decisions. They focus on results the commander expects to achieve. These OPLANs and OPORDs provide the basis for ensuring that MEDEVAC operations are comprehensively planned, synchronized, responsive, and ensure a seamless continuum of care.

EVACUATION PLANS AND ORDERS

5-10. Evacuation plans and orders help form the basis commanders use to synchronize military operations. They encourage initiative by providing the *"what"* and *"why"* of a mission and leave the how to accomplish the mission to subordinates. They give subordinates the operational and tactical freedom to accomplish the mission by providing the minimum restrictions and details necessary for synchronization and coordination.

CHARACTERISTICS

5-11. As MEDEVAC operations are complex and may cross command and Service boundaries, it is essential that the MEDEVAC annex to the medical OPLAN/OPORD clearly delineates responsibilities, communications interoperability and procedures, and coordination requirements.

PLANS AND ORDERS

5-12. Publishing the commander's OPLAN/OPORD permits subordinate commanders to prepare supporting plans and orders. They can discern and implement instructions derived from a higher commander's plan and order. Appendix B provides an example of operational planning and a sample OPORD. Additionally, the commander's OPLAN/OPORD—

- Focuses subordinates' activities.
- Provides tasks and activities, constraints, and coordinating instructions necessary for mission accomplishment.
- Encourages agility, speed, and initiative during execution.
- Conveys instructions in a standard, recognizable, clear, and simple format.

The amount of detail provided in an evacuation plan or order depends on several factors, to include the experience and competence of subordinate commanders, cohesion and tactical experience of subordinate units, and the complexity of the operation. Commanders balance these factors with their guidance and commander's intent, and determine the type of plan or order to issue. To maintain clarity and simplicity, plans and orders include annexes only when necessary and only when they pertain to the entire command. Annexes contain the details of support and synchronization necessary to accomplish the mission. For more information on the military decision-making process and development of OPLANs and OPORDs refer to ADP 5-0 and FM 6-0.

MEDICAL EVACUATION PLANNING

5-13. Medical evacuation planning which supports the AHS plan is an ongoing process and is accomplished by medical operations elements at battalion, brigade, and EAB levels. The AHS operations planning process must take under consideration all issues that could affect or influence HSS/FHP support for tactical operations. These issues should be considered in the initial developmental stages of the theater, EAB, or brigade plan as the supporting HSS appendix to the sustainment annex and the FHP appendix to the protection annex is developed.

5-14. Planning is a dynamic and continuous process. The framework established in a robust initial plan can be built on in subsequent phases of an operation. The evacuation plan for a given AO will require adjustment and change during different phases of an operation. For example, transitioning from offensive and defensive task to operations dominated by stability tasks will generally dictate a significant change in evacuation coverage priorities.

5-15. The AHS principles guide medical planners in developing OPLANs which are effective, efficient, flexible, and executable. Army Health System plans are designed to support the commander's scheme of maneuver while still retaining a focus on the delivery of medical care.

5-16. Participating in the development of the OPLAN or the OPORD ensures that the medical planner conforms to strategic, operational, and tactical plans. By participating in the orders process and developing a MEDEVAC plan, the medical planner ensures MEDEVAC support is arrayed on the battlefield in the right place, at the right time, and synchronized across operational commands to maximize responsiveness and effectiveness.

5-17. En route medical care provided during MEDEVAC must be effective and continuous to prevent interruptions in the continuum of care. Medical evacuation resources provide continuity between the roles of care within the joint operations area. They also provide interface with other deployed elements of the Military Health System operated by the other Services to enhance and facilitate the continuum of care from the POI to the CONUS-support base.

5-18. A thorough and comprehensive MEDEVAC plan is essential to establishing and maintaining control of MEDEVAC operations characterized by decentralized execution of the plan. The MEDEVAC plan complies with the CCDR's guidance and intent and maximizes the use of scarce MEDEVAC resources. Medical planners must ensure medical control is exercised over the execution of ground MEDEVAC operations and that medical influence (technical and operational supervision) is exercised over the execution of air ambulance operations. The MEDEVAC system must be responsive to changing requirements and tailored to effectively support the forces within an assigned AO. It is essential that medical control and influence be retained at the highest level of command consistent with the tactical situation.

5-19. The location of MEDEVAC assets in support of combat operations is dictated by orders and the tactical situation METT-TC. Accurately determining time and distance factors and the availability of evacuation resources are critical to determining the disposition of evacuation assets. Medical evacuation time must be minimized by the effective disposition of resources, ensuring close proximity of both supported elements and MTFs. Medical evacuation assets cannot be located so far forward that they interfere with the conduct of combat operations. Conversely, they must not be located so far to the rear that medical treatment is delayed due to lengthy evacuation routes.

5-20. Medical evacuation ground and air resources are arrayed on the battlefield to best support both the tactical commander and the AHS. It is essential that AHS assets are positioned to rapidly locate, acquire, treat, stabilize, and evacuate combat casualties. Depending upon the situation, evacuation resources may be placed in a direct support role to support maneuvering forces or area support, during operations predominated by stability tasks which are centrally located to support the mission.

5-21. Changes in tactical plans or operations may require redistribution or reallocation of MEDEVAC resources. The MEDEVAC plan must be designed to ensure flexibility and agility, as well as enhance the ability to rapidly task-organize and relocate MEDEVAC assets to meet changing battlefield requirements.

Operational and Tactical Evacuation Planning

5-22. The principle of mobility is to ensure that AHS assets remain in supporting distance to support maneuvering combat forces. The mobility, survivability (such as armor plating), and sustainability of medical units that are organic to maneuver elements must be equal to the forces being supported. Army Health System support must be continually responsive to shifting medical requirements in an operational environment. Medical evacuation must be synchronized to provide timely, responsive, and effective support to the tactical commander. The only means available to increase the mobility of medical units is to evacuate all patients they are holding. Medical units anticipating an influx of patients must MEDEVAC patients on hand prior to the start of the engagement.

5-23. Medical evacuation planning considerations, although essentially the same at all levels of command, may entail a broader scope, complexity, and detail at the higher levels of command due to the coordination and synchronization requirements which exist across command lines and service or multinational boundaries. Table 5-1 provides MEDEVAC planning considerations at various levels of command based on METT-TC.

Table 5-1. Medical evacuation planning considerations

Mission	1. Define the protocol for a valid evacuation mission (mission statement). This protocol will serve as the authority to execute the mission by each medical evacuation unit/element/crew. 2. Use the Army Health System plan and mission analysis to task evacuation units where needed. This may involve tasking part or all of direct support units to perform an area support mission. 3. Based on the type of operation— • Define trigger points for changes in the evacuation support plan. • Develop a concept of operations for all of the echelon above brigade resources. 4. Define task-organization of units assigned to the echelons above brigade and when directed to the joint task force and support relationships, if possible. Lines of coordination, and mutual support relationships should be clearly delineated. 5. Area support coverage areas will be defined with flexibility built into the plan. Coverage areas along echelons above brigade boundaries, for example, should not restrict or hinder evacuation of neighboring units that are geographically close. 6. Define pilot certifications for shore-to-ship medical air evacuation. 7. Develop standard operating procedure requirements for single versus dual-ship medical air evacuation. Address theater-specific procedures for in-flight linkup between medical evacuation aircraft and nonmedical evacuation aircraft for the purpose of dual ship requirements and/or security of the pickup zone. 8. Delineate any nonstandard air evacuation missions and allocate resources accordingly, such as support to Army special operations or joint and multinational forces. 9. Coordinate with any units that will have shared responsibilities for evacuation in a specific area and/or a specific mission.

Table 5-1. Medical evacuation planning considerations (continued)

Mission (Continued)	10. Evaluate medical evacuation specific airspace mission command and considerations. Specifically, coordinating the crossing of brigade/echelons above brigade boundaries/sectors. Medical evacuation must be afforded special expedited procedures for medical evacuation missions. Additionally, mission, enemy, terrain and weather, troops and support available, time available, civil considerations (mission variables)-dependent air corridors may have to be developed for the medical evacuation mission.
Enemy	1. Define the level or air-to-air threat and the threat faced by ground evacuation resources and the impact of this threat on the overall evacuation plan. This will be a major consideration in dictating which means of evacuation will be the primary means. 2. Based on the estimate of the number of enemy prisoners of war or other categories of detainees involved in the operation— • Define procedures for evacuating wounded, injured, or ill persons through medical channels to include responsibilities of the echelon commander for designating guards for those evacuated directly from the point of capture to a medical treatment facility. • Allocate assets as necessary to provide this support and to support detainee collection points and holding areas.
Terrain and weather	Define the types of terrain in the area of operations and use the information as a major consideration in dictating the primary means of evacuation (ground or air in each sector) and any factors that may impair evacuation efforts. For example, lines of patient drift may be identified by studying the terrain to identify the unrestricted areas likely to be used as natural corridors.
Troops and support available	1. Disseminate locations of facilities used in evacuation, medical treatment facilities, mobile aeromedical staging facilities, and aerial ports of debarkation. 2. Disseminate the capabilities and availability of joint evacuation assets and how to request their assistance. 3. Collect contact information from all evacuation units, assemble, and publish for theater use. 4. Standardize a mandatory evacuation mission data collection format, and collection method and schedule.
Time available	1. Define the acceptable limits of evacuation time based on the distance from evacuation units and casualty's evacuation precedence. 2. Determine the proper allocation of resources to support the entire area of operation requirements. 3. Plan for trigger points (in time, distance, or control measure) for changes in the evacuation plan to occur. 4. Define phases of the operation that reflect the commander's intent. Major changes in evacuation coverage usually occur when changes occur in the operational tempo. 5. Define briefing levels for launch authority. Consider developing a uniform theater wide aviation risk assessment sheet.
Civil Considerations (mission variables)	1. Based on the health service support/force health protection estimate and plan, determine requirements for the evacuation of host-nation civilians and others. 2. Define and disseminate protocols and procedures for evacuating civilians. 3. Allocate resources to support this mission, when directed. 4. Develop methods to mitigate the impact of displaced civilians to ground evacuation routes.

5-24. For an effective AHS plan, medical staffs must coordinate and share operations information in order to synchronize support operations for mission accomplishment. For additional information on HSS/FHP operations see FM 4-02, ATP 4-02.55, ATP 4-02.5 and ATP 4-02.42. After the AHS plan is completed, it is incorporated into the mission command headquarters plan. After all annexes and appendixes of the plan are approved by the commander, it is incorporated into the OPORD.

SURGEON'S RESPONSIBILITIES

5-25. The surgeon is responsible for supervision and development of AHS input for the OPORD. The AHS plan serves as the base document for this input. The AHS plan is revised or updated based on mission analysis or changes in AHS requirements. The surgeon is tasked by the assistant chief of staff, operations (G-3) and battalion or brigade operation staff officer (S-3) for AHS input to the OPORD for support of operations. The G-3/S-3 indicates timeline requirements. The surgeon and his staff are involved in all stages of the planning process, to identify all AHS requirements. Information on development of running estimates and the OPLAN/OPORD are discussed below. The medical plan/operations cell develops an AHS plan based on guidance received from the commander and the surgeon, and provides AHS operational planning updates to the surgeon. The plan is briefed to the commander for approval, as required and provided to the G-3/S-3 according to the format in ADP 6-0.

5-26. It is essential that the MEDEVAC plan for all operations be well planned, coordinated, and disseminated. In designing the MEDEVAC plan, the medical planner considers the use of CCPs, AXPs, the ambulance shuttle system, and the staffing of relay, loading, and ambulance control points of the ambulance shuttle system.

This page intentionally left blank.

Chapter 6

Medical Regulating

Medical regulating is designed to ensure the efficient and safe movement of patients. It is a system that entails identifying the patients waiting evacuation, locating the available beds, and coordinating the transportation means for movement.

PURPOSE OF MEDICAL REGULATING

6-1. Careful control of patient evacuation to appropriate MTFs is necessary to—
- Effect an even distribution of cases.
- Ensure adequate beds are available for current and anticipated needs.
- Route patients requiring specialized treatment to the appropriate MTF.

6-2. The factors that influence the scheduling of patients movement include the following:
- Patient's medical condition (stabilized to withstand evacuation).
- Tactical situation.
- Availability of evacuation means.
- Locations of MTFs with special capabilities or resources.
- Current bed status of MTFs.
- Surgical backlogs.
- Number and location of patients by diagnostic category.
- Location of airfields, seaports, and other transportation hubs.
- Communications capabilities (to include radio silence procedures).

MEDICAL REGULATING TERMINOLOGY

6-3. As medical regulating may include coordination with other services, it is necessary to use the correct terminology. These terms include—
- Intratheater patient movement—Moving patients within the theater of a combatant command or CONUS.
- Intertheater patient movement—Moving patients, into, and out of different theaters of the geographic combatant command and into CONUS or another supporting theater.
- Intracorps medical regulating—This is the system by which patients are transferred or evacuated from an MCAS/BSMC or an FST or FRST to a Role 3 hospital.
- Intratheater medical regulating—This is the system that uses theater resources to evacuate patients to and between MTFs within a theater of operations. Intratheater patient movement requires a coordinated effort between Services, hostnations MTFs, and Service component organic and theater evacuation assets.
- Intertheater medical regulating—Intertheater patient evacuation that is usually supported by the USTRANSCOM. Intertheater patient movement operations may serve as the interface between the theater and CONUS. Moving patients between, into, and out of the different theaters of the geographic combatant command and into the CONUS or another supporting theater.
- Patient administration division (PAD)—The PAD coordinates the movement of patients within and from theater as well as ensure the continuity of patient records starting at Role 2 through Role 4 treatment facilities. Medical regulating functions include consolidating all evacuation requests within the MTF and forwarding an evacuation request to the next higher headquarters

for action. The PAD is responsible for keeping the next higher echelon informed of all statuses. PAD duties include—
- Matching existing medical capabilities with reported patient needs.
- Scheduling and arranging movement of U.S., NATO, or DOD patients and authorized beneficiaries.
- Establishing procedures regulating the evacuation of patients and incoming patients.
- Tracking near real time patient care data globally, and regulate patients with a single DOD automated information system that supports medical regulating, patient movement, and patient in transit visibility.

● Medical regulating officer—The MRO maintains bed status and surgical backlogs to regulate incoming patients throughout the division theater facilities. The MRO identifies patients awaiting evacuation, locates available beds, and coordinates the means for transportation of movement while maintaining patient accountability. Under the guidance of the division surgeon or subordinate MTF, the MRO regulates and supports medical regulating, patient movement, and patient in-transit visibility requirements. MRO duties include—
- Managing what patient classes are regulated into the facility.
- Determining what resources are available to move the patients and coordinating for the use of these assets.
- Maintaining accountability of patients within the MTFs.
- Preparing reports as required.

● Theater patient movement requirements center—This is a USTRANSCOM capability assigned to manage patient movement requests within a specified geographic area or joint operational area. The TPMRC–Americas manages patient movement to, from, and within United States Northern Command and United States Southern Command. The TPMRC–East manages patient movements to, from, and within United States Central Command. The TPMRC–West manages patient movement to, from, and within the United States Pacific Command to include Antarctica. The patient movement requirements centers are responsible for theaterwide patient movement and coordinate with MTFs to identify the proper treatment and transportation component or other agencies responsible for examining the mission (see JP 4-02). The functions of the TPMRCs include—
- Consolidating requests for transfers to CONUS facilities and maintaining direct liaison with the global patient movement requirements center (GPMRC), MROs of component Services and transportation agencies which furnish evacuation transportation.
- Obtaining periodic reports of available beds from the Services MROs providing hospitalization.
- Selecting hospitals based on the reported bed availability to receive patients within the EAB.
- Maintaining direct liaison with the GPMRC, the MROs of component Services, and the transportation agencies which furnish the means for evacuation.

● Global patient movement requirements center—The GPMRC is a joint activity reporting directly to the USTRANSCOM surgeon; it serves as DOD's single manager for the development of policy and standardization of procedures and information support systems for global patient movement. The GPMRC implements policy and standardization for the regulation of clinical standards, and the safe movement of uniformed Services designated patients. The GPMRC orchestrates, and maintains global oversight of the patient movement requirements centers in coordination with the ground combatant commands geographic combatant commands and external intergovernmental organizations, as required. The GPMRC is responsible for the synchronization of current and future operational patient movement plans to identify available assets and validate transport to bed plans. (See Department of Defense instruction [DODI] 6000.11.)

● Joint patient movement requirements center (JPMRC)—A JPMRC is a joint agency established to coordinate the joint patient movement requirements function for a joint task force operating within a combatant command area of responsibility. It coordinates with the TPMRC for intratheater patient movement and the TPMRC-Americas for intertheater patient movement.

Medical Regulating

Synchronization of plans and additional guidance related to the worldwide patient movement system through the GPMRC.

- Theater Aeromedical Evacuation System (TAES)—The TAES is a functional organization provided by the USAF and performs the mission of theater aeromedical evacuation. Significant components of the TAES include the following—
  - Aeromedical evacuation command squadron.
  - Aeromedical evacuation control team (AECT).
  - Aeromedical evacuation liaison team (AELT).
  - En Route Patient Staging System.
  - Aeromedical evacuation crews.
  - Critical care air transportation teams (not necessarily USAF).
- Aeromedical evacuation command squadron—The aeromedical evacuation command squadron provides mission command of all assigned TAES forces and can deploy in advance of other aeromedical evacuation components to arrange support requirements for aeromedical evacuation forces. The command squadron advises commanders and other personnel/agencies on aeromedical evacuation operations, capabilities, and requirements and provides procedural and technical guidance for attached and transiting aeromedical evacuation elements.
- Aeromedical evacuation control team—The AECT is a cell within the air operation center and one of the core teams in the air mobility division. It is the USAF element responsible for linking validated PMRs with the appropriate airlift platform. The AECT is normally located within the air mobility division of the theater air operations center for the intratheater and the tanker airlift control center for intertheater coordination of aeromedical evacuation operations using USAF aircraft. The CONUS and strategic air operations center for the USAF is the tactical airlift control center located at Scott Air Force Base, Illinois. The AECT coordinates with and receives requirements from the patient movement requirements center for execution.
- Aeromedical evacuation liaison team—The AELT provides a direct communication link and immediate coordination between the user service and the TAES. The AELT is located at any level where USAF fixed-wing requests are initiated, verifies and coordinates with the AECT/patient movement requirements center the physiology of flight issues and patient flight/movement requirements, assists with patient preparation for flight, and directs patient on-load activities.
- En Route Patient Staging System—The ERPSS is a flexible, modular patient staging system able to operate across the full range of military operations. Utilizing the ERPSS force modules, headquarters Air Mobility Command is able to accomplish patient staging from bare base operation to a 200-bed patient staging capability at fixed facilities and allow flexibility with changing operations.
- Military Sealift Command—The Military Sealift Command is the U.S. Navy element responsible for coordinating movement of supplies, equipment, and personnel into the AO by Navy ships. Further, it coordinates, through the TPMRC, the MEDEVAC of patients by ship from the AO to the support base or hospital ship, as required.
- Corps movement control center—The corps movement control center is the corps movement control organization. It provides centralized movement control and highway regulation for movement of personnel and material into, within, and out of the corps area. When USAF capabilities are exceeded, the corps movement control center coordinates requests for additional air and ground resources. It also obtains the necessary clearance to support the MEDEVAC mission from the operations area.
- Theater Army movement control agency—The theater Army movement control agency mission is to provide movement management services and highway traffic regulations and to coordinate for personnel and material movements into, within, and out of the theater. The theater Army movement control agency coordinates with multinational and host-nation movement control agencies. It also coordinates with the USTRANSCOM and its subordinate units (such as the Air Mobility Command and Military Sealift Command) and prepared movement and port clearance plans and programs.

- Joint military transportation board—The joint military transportation board is a joint staff composed of members of the Army, USAF, and United States Navy that coordinates transportation requirements for patients requiring intertheater evacuation.
- United States Transportation Command Regulating and Command and Control Evacuation System—The USTRANSCOM Regulating and Command and Control Evacuation System is the single DOD automated information system that supports medical regulating, patient movement, and patient in-transit visibility requirements.

MEDICAL REGULATING FOR THE ECHELONS ABOVE BRIGADE

6-4. Medical regulating in and from the division is the responsibility of the division surgeon section. Medical regulation in the division is not as formalized as the rest of the medical regulating system. It is usually operated procedurally so as not to depend solely on communications to effect rapid evacuation. The medical regulating function in the division surgeon section is concerned primarily with—
- Tracking the movement of patients throughout the division MTFs and into the theater facilities.
- Monitoring the use of ambulance assets.

6-5. Air and ground ambulances placed in support of the BCTs can be positioned in the BCT support area when the mission directs. When these assets go forward to the BSMC or the MCAS to evacuate patients to EAB MTFs, they have MTF destinations predetermined (blocks of beds). The division surgeon section, in coordination with the MEDBDE (SPT) MRO, establishes the number of patients a supporting MTF can accept during a particular period of time. These blocks of available beds are then provided to the supporting ambulances prior to the call for missions. Additional requirements include—
- Once an evacuation mission is completed, the originating division MTF provides the patient disposition to the owning brigade surgeon section—
    - Patient numbers by category and precedence.
    - Departure times.
    - Modes of transportation.
    - Destination MTFs.
    - Any other information required by SOP.
- The division surgeon section, in turn, notifies the MEDBDE (SPT) MRO via the appropriate patient administration net, which is monitored by the EAB MTFs. Air ambulance and EAB ground ambulances and can relay patient information to the receiving MTF.
- To reduce the turnaround time for ground ambulances and to move more serious patients to Role 3 hospitals in the support areas—
    - Air ambulances are given blocks of beds in the MTFs farther to the rear.
    - Ground ambulances are normally given blocks of beds in the more forward deployed Role 3 hospitals.

6-6. Medical evacuation can be effected immediately, procedurally, and under conditions of communication silence without interrupting the continuum of care by—
- Preparing patient estimates.
- Prioritizing and task-organizing ambulance support.
- Assistant chief of staff (personnel) is also involved in medical regulating as the assistant chief of staff (personnel) is responsible for personnel accountability and casualty reporting. Casualty liaison teams will normally be collocated at Role 3 hospitals to assist and coordinate patient tracking and casualty reporting.

MEDICAL REGULATING WITHIN THE OPERATIONAL AREA

6-7. The requirement to transfer patients from one MTF to another MTF within the operational area occurs. This results from—
- Surgical backlogs.
- Mass casualty situations.
- Specialty care requirements.
- Planned movement of an MTF.

6-8. When it is necessary to transfer a patient, the attending physician notifies the MTF PAD. The PAD consolidates all such requests from the MTF and works with the MRO to request movement authority from the MEDBDE (SPT) patient movement branch.

6-9. If the medical patient movement branch can transfer the patient or patients to its subordinate MTFs, the MRO at the patient movement branch designates the MTFs to receive the patients and notifies both the requesting and receiving MTFs of the transfer. The MRO also tasks subordinate MEDEVAC units for the assets to transfer the patients.

6-10. If the MEDBDE (SPT) cannot provide the needed hospitalization within its own resources, the MRO forwards the request to the medical TPMC for action. The TPMC then designates the receiving MTFs and notifies the subordinate patient movement branch. The MEDBDE (SPT) patient movement branch disseminates the information to the hospital, Role 3, PAD and coordinates the evacuation resources for the transfer. The MRO can also coordinate the regulation of patients to—
- Other U.S. military Service MTFs and naval hospital ships.
- Allied nations' military hospitals.
- Other authorized supporting facilities.

MEDICAL REGULATING FROM THE OPERATIONAL AREA TO ECHELONS ABOVE BRIGADE

6-11. Medical treatment facilities attending physicians and oral and maxillofacial surgeons submit daily reports to the hospital PAD listing the patients requiring MEDEVAC or transfer. The PAD assembles this information and transmits the report to the MEDBDE (SPT) headquarters. This report is a request for transportation, as well as a notification of the number of patients requiring evacuation. The report classifies the patients according to—
- Diagnostic category.
- Desired on-load points.
- When the patients will be available for evacuation.

6-12. The MEDBDE (SPT) patient movement branch consolidates these reports from each hospital attached to the MEDBDE (SPT) and forwards the data to the TPMC.

6-13. If a JPMRC has been activated within the theater, the MEDCOM (DS) MRO consolidates all reports from the MEDBDE (SPT) and forwards them to the JPMRC. The JPMRC designates the MTFs in the EAB to receive the patients. The designation is based on the previously received bed status reports from all Service components and available means of evacuation. The JPMRC then notifies the MEDCOM (DS) MRO of the designated MTF. The TPMC accomplishes this task if the JPMRC is not activated.

6-14. The primary means of moving patients from the AO out of theater is USAF aircraft missioned to perform aeromedical evacuation. With the elements of the TAES deployed, it is possible to find AELTs at each role and as far forward as USAF fixed-wing requests are initiated. The AELT monitors the MRO patient evacuation requests and passes requirements to the TPMRC. At the same time they pass airlift requirements to the AECT, seeking an aircraft to perform the evacuation mission. The AELT, through the MRO patient movement request, requests the TPMRC to move patients. Included in the request are the originating medical facility and destination airfields.

6-15. The AECT is a component of the TAES and performs the mission of coordinating the movement of and providing in-flight medical care to patients while under the USAF control. The AECT receives patient movement requirements from the TPMRC, then works with the airlift control team in the air operations center to meet the evacuation requirement.

6-16. The air operations center coordinates the forward movement of cargo and personnel aboard USAF aircraft with other USAF units, Army transportation representatives, and United States Navy agencies. Some of these aircraft are scheduled to evacuate patients on their return trips. These aircraft seldom go forward solely to evacuate patients; however, these missions may be used for retrograde airlift and mixed cargo/aeromedical evacuation missions is the primary means of aeromedical evacuation due to limited aircraft and the need to maximize the use of these assets.

6-17. After the schedules have been arranged, the AECT returns the detailed flight schedule to the JPMRC who passes the information to the theater TPMC/AELT and the parent aeromedical evacuation element.

6-18. The MEDCOM (DS) MRO, in coordination with the AELT, issues these instructions to the patient movement branches (with the authority to move patients in the AO Army MTFs) and the receiving MTFs. The MTFs must prepare to receive the patients at the destination airfields (this may be accomplished by collocating an MCAS at the airfield to receive the incoming patients). The patients are sorted by destination hospital and moved by Army MEDEVAC means. The instructions mentioned above include, as a minimum, the—

- Numbers of patients to be moved.
- On-load airfields.
- Destination airfield.
- USAF aircraft mission number.
- Estimated time of arrival at the destination airfield.

6-19. The TPMC issues the flight and movement instruction to its subordinate medical patient movement branches. The patient movement branch then directs the evacuation units and hospitals within their AOs to move the patients to the on-load airfield according to the arrival time of the aircraft. This movement must be closely controlled; the length of stay in an ERPSS facility may be from 2 to 72 hours. Holding times differ depending on the size of the staging facility. At the 10-bed initial stage, holding times are limited by the amount of space and supplies in this initial package. Two to four hours is the preferred hold time; not to exceed six hours. At larger staging facilities (50 to 100 bed), holding times will vary depending on the operational capability of the location and the flight schedules, but should be limited to no more than 72 hours.

MEDICAL REGULATING WITHIN ECHELONS ABOVE BRIGADE

6-20. Medical regulating within the EAB is similar to the system used within the operational area. Attending physicians within the Role 3 MTFs notify the hospital PAD of patients requiring evacuation to Role 3 hospitals. The PAD then consolidates the requests from the hospital and forwards the consolidated requests to the TPMRC for intertheater aeromedical patient movement.

6-21. The TPMC, based on periodic bed status and availability reports from subordinate MTFs, designates specific hospitals to receive the patients. The MTFs are designated based on bed availability, to include specialty beds, to support the specific patients. The TPMC then notifies the requesting medical patient movement branch of the designated MTFs and, in turn, notifies the designated MTF.

6-22. The MROs track the availability of medical specialties at each MTF and the specific number of beds that can be assigned to that specialty. A bed report is given for the medical capabilities available at each MTF which helps designate which hospital a patient will be evacuated to. Bed reporting is broken down into Contingency Bed Reporting and Noncontingency Bed Reporting. The requesting provider will call to verify that the receiving provider has accepted the patient and is aware the patient is inbound.

INTERTHEATER MEDICAL REGULATING

6-23. The patients who are evacuated to EAB are treated there and then further evacuated to CONUS or another supporting theater. The attending physicians at the MTF notify the PAD. The PAD then consolidates these requests and forwards them to the patient movement branch. This patient movement branch forwards the consolidated request to the TPMC that, in turn, consolidates and forwards a request to the JPMRC (if established) or the TPMRC.

6-24. Upon request of the JPMRC/TPMRC for authority to evacuate patients to CONUS, the global patient movement integration center directs the distribution of these patients into MTFs throughout CONUS, advises the JPMRC/TPMRC of the destination MTF; and provides the authority for such movement. As a rule, the designated MTF is a military facility. Civilian national disaster medical system member hospitals and other federal hospitals may also receive patients. The Department of Veterans Affairs hospitals and other federal hospitals may also receive patients who are expected to be discharged from Service. The TPMRC validates the patient movement requirements and, if moving by air, tasks the tactical airlift control center to plan, schedule, and execute the intertheater evacuation.

6-25. When the JPMRC/TPMRC receives the authorization to move patients, it notifies the TPMC of destination hospitals in CONUS. The TPMC coordinates with the joint military transportation board to arrange movement of CONUS-bound patients. The TPMC then authorizes the movement to aeromedical staging facilities that are located on or near air bases or airstrips capable of handling long-range aircraft. Transportation is arranged, within Army channels, to move patients from the hospitals to the staging facilities. The MEDBDE (SPT), in coordination with AELT, then notifies the subordinate Role 3 hospitals of the flight schedule and the evacuation arrangements for movement to the strategic airheads. At strategic airheads, there is an established aeromedical staging facility. When the patients are delivered to the USAF, the responsibility for those patients is transferred from the Army MTF to the TAES. Upon arrival in CONUS further movement is the responsibility of the TPMRC-Americas.

6-26. All patients may not be able to be moved by air from the theater to CONUS. In that event, the Military Sealift Command is used to move them by surface means. The movement authority also comes from the TPMRC/JPMRC or TPMC which has arranged with the Navy Service Component Command for the movement of patients by hospital ships. When the patients are moved by ships, the MEDCOM (DS) has to provide holding facilities at the port (collocating an MCAS can provide this support). Patients are delivered to these holding facilities and held there until loaded aboard the ships.

6-27. Routine Flight Evacuation. Patients who are evacuated to CONUS via routine air and are generally not patient movement request flights are prearranged by the Service member's unit for redeployment. These patients have medical conditions that do not allow them to remain, do not need en route care, and are not authorized aeromedical evacuation.

This paragraph implements ABCA Standard 2080

6-28. This agreement applies solely to the unclassified United States Transportation Command regulating and Command and control Evacuation system (TRAC2ES) information technology system. The U.S. TRAC2ES is a patient movement automated information system that assembles, assesses, and prioritizes patient movement requirements. This system automates the aeromedical evacuation and medical regulation to provide patient visibility and accountability in-theater and out of theater to home country. This capability is granted to ABCA nations, limited to the TRAC2ES-Web (T-Web) and TRAC2ES-Mobile (T-Mobile) subsystems. Other capabilities of TRAC2ES may be requested from U.S.TRANSCOM as required.

EN ROUTE PATIENT STAGING SYSTEM

6-29. Aeromedical evacuation is necessary for patients who require definitive care, not available in theater Role 3 MTFs, or when the patient's condition exceeds the theater evacuation policy. Patient staging is a key component of ensuring timely aeromedical evacuation. Patient staging is designed to temporarily stage patients and prepare them for aircraft loading while reducing the amount of time an aeromedical evacuation aircraft is on the ground.

Chapter 6

6-30. This staging system is used to evacuate patients from—
- United States Air Force operational locations within the operational area to MTFs outside the operational area.
- Airheads or airborne objective areas where airborne operations include USAF forward logistics support.

6-31. Historically patient staging has been accomplished via mobile aeromedical staging facility or contingency aeromedical staging facility. The USAF has transitioned from the mobile aeromedical staging facility and contingency aeromedical facility and developed the modular building block approach to achieve any size patient staging facility requirement called the ERPSS. The ERPSS coordinates and communicates with medical and aeromedical evacuation elements to accomplish patient care, staging and patient movement.

6-32. The ERPSS has a two-fold mission—to provide support and continuity of medical care for patient movement and serve as an integral patient interface to the en route care system. The ERPSS provides personnel and equipment necessary for 24-hour patient staging operations, patient transportation between the staging facility and the aircraft, and administrative processes for tracking patients transiting the en route care system worldwide.

6-33. The ERPSS has no surgical, laboratory, dental, mental health, x-ray, or blood bank capabilities. Therefore, it is normally collocated with an MTF capable of providing required inpatient and outpatient services. Critically ill patients and inpatient psychiatric patients must be staged/held at the collocated or supporting MTF. The collocated or supporting inpatient MTF is responsible for holding/staging all patients with medical care requirements outside the scope of the ERPSS.

6-34. The ERPSS is designed to temporarily hold patients as they transit the aeromedical evacuation system while providing short-term complex medical-surgical nursing care and limited emergent intervention. The length of stay in an ERPSS facility may be from 2 to 72 hours. Holding times differ depending on the size and location of the staging facility. At the 10-bed initial stage, holding times are limited by the amount of space and supplies in this initial package. Two to four hours is the preferred hold time; not to exceed six hours. At the larger staging facilities (50 to 100 bed), holding times will vary depending on the operational capability of the location and the flight schedules, but should be limited to no more than 72 hours.

6-35. All ERPSS facilities (with the exception of the base unit, the tactical ERPSS 10) should be collocated with a bedded facility in order to maximize its capability to care for patients. Primary medical/surgical and other ancillary services must be available 24 hours a day for patients in the staging facility. If the staging facility is not collocated with an MTF that can provide the required clinical support, arrangements must be in place to meet the clinical support levels needed for patients transiting the facility.

6-36. En route patient staging system is a medical war reserve materiel supply and equipment asset, with a building construct that allows medical planners to right size the facility requirements based on patient movement requirements. The ERPSS is able to build incrementally from a 10-bed mobile facility to a fixed 200-bed staging facility. This is accomplished as workload changes, or is projected to change, unit type code packages (personnel and equipment) may be deployed in small increments and combined with previously deployed ERPSS unit type codes to increase capability. For more information on the ERPSS refer to AFTTP 3-42.57.

LIMITATION OF THE UNITED STATES AIR FORCE THEATER AEROMEDICAL EVACUATION SYSTEM

6-37. There are a number of limitations that are inherent in the current system. These include the following:
- No organic CBRN decontamination capability.
- The ERPSS does not have the capability to provide patient meals.
- All ERPSS facilities (with the exception of the ERPSS-10) should be collocated with an MTF to support patient treatment and movement.

- The EPRSS initially deploys with a limited number of days of supplies and is reliant on reach back and supporting MTFs for logistical support until a mature logistical system is developed.
- It is the Army's responsibility to provide food and other logistical support required including moving patients back to Army facilities should USAF aeromedical evacuation support be delayed.

ORIGINATING MEDICAL FACILITY'S RESPONSIBILITIES

6-38. Once the authorization to move the patient has been given, the originating medical facility must complete the following administrative procedures prior to entering the patient into the TAES:

- The patient's baggage tag, patient evacuation manifest, and patient evacuation tag are the specified evacuation forms for all services and are completed as required by disservice regulation.

Note. If the originating medical facility is a Role 2 facility (such as the BSMC), the forms may not be available, in which case, the forms will need to be completed by the ERPSS.

- All of the patient's medical records must be collected together and packaged. The dental records are forwarded separately in the event they are needed for identification.
- At the appropriate time, the originating medical facility provides transportation to the ERPSS and assists in the off-load.
- The originating medical facility must provide the necessary medications, medical supplies, and equipment to support the patients' travel times to regulated destination.
- Any requirements for armed guards must be met by the echelon commander.

Note. Medical units do not provide guards for prisoners or detainees in their care. When guards are required, they are provided by the echelon commander. The originating medical facility will coordinate for this support when needed.

- A limited amount of personal baggage is authorized if each piece is properly tagged and delivered to the ERPSS with the patient. Dependent on the threat, patients should always be evacuated with CBRN protective equipment.
- Each patient must be clearly identified with a wristband or equivalent identification and properly classified as to their medical condition.
- The originating medical facility should ensure that each patient is properly briefed and prepared for their evacuation prior to their arrival at the ERPSS.

MEDICAL REGULATING FOR ARMY SPECIAL OPERATIONS FORCES

6-39. As in MEDEVAC, the medical regulating plan must be integrated with the ARSOF operational and logistics plan. Maximum use of opportune (operational and logistics) aircraft and command and logistics communications nets must be coordinated to expedite mission requests and ensure success

6-40. The ARSOF medical planner must constantly coordinate with the battalion or group operations and logistics sections to obtain up-to-date information of opportune transportation assets to be used for evacuation. In an operation, or when the theater is sufficiently developed to allow the TAES to be used effectively, the primary means of air evacuation will be those special operations aviation or USAF SOF airframes conducting the clandestine mission. It is essential that coordination is made through the theater special operations command or the mission command element for SOF medical or pararescuemen to accompany the flight when backhauling the SOF patient. Otherwise, a medic from the SOF unit being supported may have to accompany the patient, leaving the mission without proper medical support, or the patient may have to be transported without a SOF escort to monitor and provide en route care.

6-41. For all other special operations, the supporting MEDEVAC unit provides air and ground ambulances according to standard procedures. United States Air Force ERPSS or AELTs may be collocated at

Chapter 6

SOF-support bases, or base camps, particularly during contingency operations where the build-up phase allows for pre-positioning of assets.

6-42. During sustained special operations missions, the theater special operations command cannot afford to lose the services of low-density ARSOF skilled Soldiers who become casualties. Every effort must be made to preserve ARSOF capabilities and combat power in theater. A determination must be made as to who can be treated and returned to duty at MTFs within the theater. As an exception to the theater evacuation policy, the CCDR may retain injured or wounded ARSOF in theater where they can be returned to limited duty without jeopardizing their recovery and health. There they can assume the support duties performed by other ARSOF Soldiers, freeing the latter for operational duties.

Appendix A
Geneva Conventions and the Law of War

The conduct of armed hostilities on land is regulated by customs and international law and lawmaking treaties such as The Hague and Geneva Conventions. The rights and duties set forth in the Conventions are part of the supreme Law of the Land. The U.S. is obligated to adhere to these obligations even when an opponent does not. It is a DOD and Army policy to conduct operations in a manner consistent with these obligations. An in-depth discussion of the provisions applicable to medical units and personnel is provided in FM 4-02. This appendix discusses only those articles or actions which affect MEDEVAC operations.

DISTINCTIVE MARKINGS AND CAMOUFLAGE OF MEDICAL FACILITIES AND EVACUATION PLATFORMS

This section implements STANAG 2931.

A-1. All U.S. medical facilities and units, except veterinary, display the distinctive flag of the Geneva Conventions. This flag consists of a Red Cross on a white background. It is displayed over the unit or facility and in other places as necessary to adequately identify the unit or facility as a medical facility.

AUTHORIZED EMBLEMS

A-2. The Geneva Conventions authorizes the use of the following distinctive emblems on a white background: Red Cross, Red Crescent, and Red Crystal. In operations conducted in countries using an emblem other than the Red Cross on a white background, U.S. Soldiers must be made aware of the different official emblems. Although not specifically authorized as a symbol in lieu of the Red Cross, enemies of Israel in the past wars have recognized the Red Star of David and have afforded it the same respect as the Red Cross. This showed compliance with the general rule that the wounded and sick must be respected and protected when they are recognized as such, even when not properly marked. In December 2005 during the Diplomatic Conference in Geneva, the body of the International Committee of the Red Cross adopted Protocol III to the Geneva Conventions, creating an additional emblem alongside the Red Cross and Red Crescent. The new emblem, known as the Red Crystal, resolves several issues that the International Committee of the Red Cross has faced over the years and provides an additional or third emblem devoid of any national, political, or religious connotation. This allows the countries unwilling to adopt the Red Cross or the Red Crescent to join the movement as a full member by using the Red Crystal.

A-3. United States' forces are legally entitled to only display the Red Cross. However, commanders have authorized the display of both Red Cross and Red Crescent to accommodate host nations concerns and to ensure that confusion of emblems would not occur. Such use of the Red Crescent must be in smaller size than the Red Cross.

CAMOUFLAGE OF THE DISTINCTIVE EMBLEM

A-4. Camouflage of medical facilities (medical units, medical vehicles, and medical aircraft on the ground) is authorized when the lack of camouflage might compromise tactical operations. The marking of facilities and the use of camouflage are incompatible and should not be undertaken concurrently.

Appendix A

A-5. If failure to camouflage endangers or compromises tactical operations, the camouflage of medical facilities may be ordered by a NATO commander of at least brigade-level or equivalent. Such an order is temporary and local in nature and is rescinded as soon as circumstances permit.

> *Note.* There is no such thing as a *camouflaged* Red Cross. When camouflaging a medical unit or ambulance, either cover up the Red Cross or take it down. A *black cross* on an olive drab or any other background is not a symbol recognized under the Geneva Conventions.

MEDICAL AIRCRAFT

A-6. Medical aircraft exclusively employed for the removal of wounded and sick and for the transport of medical personnel and equipment shall not be attacked, but shall be respected by belligerents, while flying at heights, times, and on routes specifically agreed upon between the belligerents concerned.

A-7. Known medical aircraft, when performing their humanitarian functions, must be respected and protected. Such aircraft does not constitute a military objective. The aircraft shall not be deliberately attacked or fired upon if identified as a protected medical aircraft.

A-8. The medical aircraft shall bear, clearly marked, the distinctive emblem together with their national colors on their lower, upper, and lateral surfaces.

A-9. Unless agreed otherwise, flights over enemy-occupied territory are prohibited.

A-10. Medical aircraft shall obey every summons to land. In the event that a landing is thus imposed, the aircraft with its occupants may continue its flight after examination, if any.

A-11. In the event that an air ambulance involuntary lands in enemy or enemy-occupied territory, the wounded and sick, as well as the crew of the aircraft, shall be prisoners of war; medical personnel will be treated as prescribed in the Conventions.

SELF-DEFENSE AND DEFENSE OF PATIENTS

A-12. When engaging in MEDEVAC operations, medical personnel are entitled to defend themselves and their patients. They are only permitted to use individual small arms.

A-13. The mounting or use of offensive weapons on dedicated MEDEVAC vehicles and aircraft jeopardizes the protections afforded by the Geneva Conventions. These offensive weapons may include but are not limited to machine guns, grenade launchers, hand grenades, and light antitank weapons.

A-14. Medical personnel are only permitted to fire in their personal defense and for the protection of the wounded and sick in their charge against marauders and other persons violating the law of war.

A-15. Medical personnel are not authorized crew-served or offensive weapons. They carry small arms, such as rifles, pistols, squad automatic weapons, or authorized substitutes in the defense of medical facilities, equipment, and personnel/patients without surrendering the protections afforded by the Geneva Conventions.

DETAINEES

A-16. Sick, injured, and wounded detainees are treated and evacuated through normal medical channels but are physically segregated from U.S. and multinational patients.

A-17. Personnel resources to guard detainee patients are provided by the supported commander. Medical personnel do not guard detainee patients.

COMPLIANCE WITH THE GENEVA CONVENTIONS

A-18. The U.S. is a party to the 1949 Geneva Conventions. Two of these Conventions afford protection for medical personnel, facilities, and evacuation platforms (to include aircraft on the ground). All AHS personnel should thoroughly understand the provisions of the Geneva Conventions that apply to medical activities.

CONSEQUENCES OF VIOLATIONS

A-19. Violation of these Conventions can result in the loss of protection afforded by them. Medical personnel should inform the tactical commander of the consequences of violating the provisions of these Conventions. The consequences may include—
- Medical evacuation assets subjected to attack and destruction by enemy.
- Health service support capability degraded.
- Captured medical personnel becoming prisoners of war rather than retained persons. They may not be permitted to treat their fellow prisoners.
- Loss of protected status for medical unit, personnel, or evacuation platforms (to include aircraft on the ground).

PERCEPTION OF IMPROPRIETY

A-20. Since even the perception of impropriety can be detrimental to the mission and U.S. interests, AHS commanders must ensure they do not give the impression of impropriety in the conduct of MEDEVAC operations. For example, if a medical commander included in the unit's SOP rules governing the use of automatic or crew-served weapons, it would give the impression that the unit possessed and intended to use these types of weapons. Under the provisions of the Geneva Conventions, medical units are only authorized individual small arms for use in the defense of the patients under their care and for themselves. Even though the unit does not possess these types of weapons, the entry in the SOP could be misinterpreted and a case made that the commander intended to use these weapons in violation of the Geneva Conventions.

This page intentionally left blank.

Appendix B
Example of the Medical Evacuation Plan and Operations Order

This appendix provides an example of a medical evacuation plan as part of an operations order.

EXAMPLE OF THE MEDICAL EVACUATION PLAN

B-1. Medical support is described in Appendix 3, Army Health System Support of Annex F, and Sustainment. Medical evacuation support including the theater evacuation policy, en route care, medical regulating, CASEVAC, and the MEDEVAC of CBRN casualties is described in the AHS Support subparagraph. Refer to FM 6-0 for further information on developing annexes and appendixes for plans and orders. See Figure B-1 for an example of the MEDEVAC plan.

(CLASSIFICATION)

Place the classification on the top and bottom of each page of the attachments. Place the classification marking (TS), (S), (C), or (U) at the front of each paragraph and subparagraph in parentheses. Refer to Army Regulation 380-5 for classification and release markings instructions.

Copy ## of ## copies
Issuing headquarters
Place of issue
Date-time group of signature
Message reference number

Include the full heading if attachment is distributed separately from the base order or higher-level attachment.

TAB C (MEDICAL EVACUATION) TO APPENDIX 3 (ARMY HEALTH SYSTEM SUPPORT) ANNEX F (SUSTAINMENT) TO OPERATION PLAN/ORDER (number) (code name)—(issuing headquarters) (classification of title)

(U) References: *List documents essential to understanding the appendix.*

   a. List maps and charts first. Map entries include series number, country, sheet names or numbers, edition, and scale.

   b. List other references in subparagraphs labels as shown

   c. Doctrinal references for medical evacuation include FM 4-02 and applicable Army Medical Department ATPs.

(U) Time Zone Used Throughout the Operation Plan (OPLAN): *The time zone used throughout the OPLAN (including attachments) is the time zone applicable to the operation. Operations across several time zones use Universal Time (ZULU) time.*

(page number)
(CLASSIFICATION)

Figure B-1. Example of a medical evacuation plan

Appendix B

(CLASSIFICATION)

1. (U) Situation. *(State the general medical evacuation factors affecting support of the operation. Include any information essential to understanding the current situation as it influences medical evacuation.)*

    a. (U) Area of Interest. *Describe the area of interest as it relates to the Army Health System and medical evacuation. Refer to Annex B (Intelligence) as required.*

    b. (U) Area of Operations. *Refer to Appendix 2 (Operation Overlay) to Annex C (Operations) as required.*

        (1) (U) Terrain. *Describe the aspects of terrain that impact medical evacuation operations. Refer to Annex B (Intelligence) as required. This paragraph should discuss any aspects of the terrain that will either hinder or enhance the execution of the evacuation mission. It should discuss both natural and man-made terrain, as medical evacuation urban areas can pose significant challenges not found on a natural battlefield.*

        (2) (U) Weather. *Describe the aspects of weather that impact on medical evacuation operations. Refer to Annex B (Intelligence) as required. This should include a discussion of current weather conditions and seasonal variants. Weather conditions impact both ground and air evacuation operations; however, the most significant impact may be on air ambulances as severely inclement weather can ground all aircraft. It should also discuss the impact that the weather has on the terrain (such as rivers being frozen in winter or tundra becoming impassable in spring). The climate may pose problems with acclimation as well as place additional requirements to sustain personnel during evacuation on evacuation assets (litter evacuation in the mountains in extreme cold weather operations may require warming tents and the capability to sustain the patient during nighttime when evacuation is difficult).*

    c. (U) Enemy Forces. *Refer to Annex B (Intelligence) as required. This section is used to list information about the composition, disposition, location, movements, estimated strengths, and identification of enemy forces. Include enemy capabilities as well as weapons systems that could influence the medical evacuation mission, if conducting operations focused on stability and defense support of civil authorities, change the title of this subparagraph to "Terrorist/Criminal Threats."*

    d. (U) Friendly Forces. *Outline the higher headquarters' medical plan. List any pertinent information concerning friendly forces that might influence the medical evacuation mission. List the designation, location and outline of plan of higher, adjacent, and other medical assets that support or impact the issuing of headquarters or required coordination and additional support. The below subparagraphs are examples of possible topics.*

    e. (U) Interagency, Intergovernmental, and Nongovernmental Organizations. *Identify and describe other organizations in the area of operations that may impact the conduct of the medical evacuation mission and operations. Refer to Annex V (Interagency Coordination) as required.*

    f. (U) Civil Considerations. *Describe the aspects of the civil situation that impact medical evacuation operations. Refer to Annex B (Intelligence) and Annex K (Civil Affairs Operations) as required. This may include the effects of dislocated civilian population and detainees have on evacuation routes and mission workload. Injured, ill, or wounded detainees are evacuated using the same evacuation means but are segregated from U.S., or multinational patients. Coordination for nonmedical guards for detainee patients being evacuated through medical channels must be accomplished with the echelon commander. A determination of eligible beneficiaries for evacuation should also be developed and disseminated.*

    g. (U) Attachments and Detachments. *List units attached or detached only as necessary to clarify task organization. Refer to Annex A (Task Organization) as required.*

    h. (U) Assumptions. *List any medical or medical evacuation assumptions that support the appendix development.*

(page number)
(CLASSIFICATION)

Figure B-1. Example of a medical evacuation plan (continued)

(CLASSIFICATION)

2. (U) **Mission.** State the mission of medical evacuation in support of the base order or plan.

3. (U) **Execution.**

    a. (U) Scheme of Medical Evacuation Support. The scheme of support describes how the commander sees the actions of subordinate units fitting together to accomplish the mission. Commanders ensure that their scheme of support is consistent with their intent and that of the next two higher headquarters. The scheme of support describes any other details the commander considers appropriate to clarify the concept of operations and ensure unity of effort. If the integration and coordination are too lengthy for this paragraph, they are addressed in the appropriate annexes. When an operation involves two or more clearly distinct and separate phases, the scheme of support may be prepared in subparagraphs describing each phase.

    b. (U) Task to Subordinate Units. List medical evacuation tasks assigned to specific units not contained in the base order.

    c. (U) Coordination Instructions. List only instruction applicable to two or more subordinate units not covered in the base order.

        (1) (U) Evacuation Routes/Corridors. Evacuation routes should be preplanned, indicated on the medical evacuation overlay, and reconnaissance accomplished. Routes that provide lucrative targets, have significant obstacles to circumvent, or will be unduly congested due to fleeing dislocated persons should only be used if no other routes are available. Air corridors are provided by the supporting airspace control order. Due the placement of hospitalization resources within the joint operations area, air ambulance may be required to operate out of fixed sectors and coordination for air escorts must be coordinated. Evacuation overlays must be developed to facilitate the evacuation effort. Both supporting and supported units must maintain and update (as required) overlays throughout the operation.

        (2) (U) Medical treatment facility Locations. List existing and proposed locations of all available MTFs.

        (3) (U) Casualty Collecting Points. Location, staffing, and activation trigger (such as crossing a phase line) must be known to supported and supporting units.

        (4) (U) Ambulance Exchange Points. Location, staffing, and activation trigger (such as crossing a phase line) must be known to supported and supporting units.

        (5) (U) Ambulance Shuttle System. The ambulance (or litter) shuttle system is a management tool to facilitate the medical evacuation of forward areas.

    d. (U) Tasks to Subordinate Units. List medical evacuation tasks assigned to specific subordinate units not contained in the base order.

    e. (U) Resources available.

        (1) (U) Supplies and Equipment. This includes both medical and nonmedical supplies and equipment. Litter and blanket exchanges should be covered here. If patient movement items (PMI) is an issue that should be discussed here. This may also include a list of equipment and resources such as the need for any special equipment to recover and move patients (special equipment for use in urban areas for extracting casualties from buildings and rubble).

        (2) (U) Additional Medical Assets. List the available medical assets from higher levels such as treatment teams, ambulances, and nonstandard vehicles available for patient movement.

(page number)
(CLASSIFICATION)

Figure B-1. Example of a medical evacuation plan (continued)

> **(CLASSIFICATION)**
>
> 4. (U) <u>Sustainment</u>.
>
>   a. (U) <u>Material and Services.</u> *(Refer to standard operating procedure (SOP) or another annex whenever practical.)*
>
>   (1) (U) <u>General supply.</u> *(Provide special instructions applicable to medical evacuation units.)*
>
>   (2) (U) <u>Classes of Supply</u>. Consider supply levels for all classes of supply, in the event of mission requirements in an austere environment and at extended distances from the full complement of logistics and sustainment resources.
>
>   b. (U) <u>Medical Logistics.</u> *Provide special procedures applicable to the operations for Class VIII resupply and medical maintenance request and procedures if different from Appendix 3 (Army Health System Support) to Annex F (Sustainment).*
>
>   (1) (U) <u>Distribution.</u> *Include the method of distribution and any limitations or restrictions that are applicable. Additionally, if special transportation requirements exist, they should also be noted.*
>
>   (2) (U) <u>Medical Logistic Activities.</u> *This includes the location of the medical supply activity supporting the area of operation (AO) and the means of communicating requests for resupply.*
>
>   c. (U) <u>Army Health Service Support</u>. *Identify availability, priorities, and instructions for medical care. Refer to Annex F (Sustainment) as required. Prescribe the plan for air and ground medical evacuation.*
>
>   (1) (U) <u>Time or Condition When a Plan Becomes Effective.</u> *Include the time or the conditions under which the plan is to be placed in effect.*
>
>   (2) (U) <u>Commander's Critical Information Requirements.</u> *List only in coordinating instructions and not in annexes.*
>
> 5. (U) <u>Command and Signal</u>.
>
>   a. (U) <u>Command</u>.
>
>   (1) (U) <u>Location of Commander</u>. *State the location of key medical evacuation commanders.*
>
>   (2) (U) <u>Succession of Command</u>. *Identify the chain of command if not addressed in unit SOPs.*
>
>   (3) (U) <u>Liaison Requirements</u>. *State the medical evacuation liaison requirements not covered in unit SOPs.*
>
>   b. (U) <u>Control.</u>
>
>   (1) <u>Command Posts</u>. *Describe the employment of medical evacuation command posts, including the location of each command post and its time of opening and closing. Include map coordinates for the command post locations and at least one future location for each command post.*
>
>   (2) (U) <u>Reports</u>. *List reports requiring special emphasis that is not addressed elsewhere. Describe medical evacuation reporting requirements. Refer to Annex H (Signal) as required.*
>
>   c. (U) <u>Signal.</u> *Address any medical evacuation-specific communications requirements. Refer to Annex H (Signal) as required.*
>
> (page number)
> **(CLASSIFICATION)**

Figure B 1. Example of a medical evacuation plan (continued)

Example of the Medical Evacuation Plan and Operations Order

(CLASSIFICATION)

ACKNOWLEDGE: Include only if attachment is distributed separately from base order.

(Commander's last name) (Commander's rank)

The commander or authorized representative signs the original copy of the attachment. If the representative signs the original, add the phrase "For the Commander." The signed copy is the historical copy and remains in the headquarters' file.

OFFICIAL:

(Authenticator's name)
(Authenticator's position)

(page number)
(CLASSIFICATION)

Figure B-1. Example of a medical evacuation plan (continued)

EXAMPLE FORMAT FOR AN OPERATIONS ORDER

B-2. Figure B-2 provides an example of medical evacuation operation order.

(CLASSIFICATION)

Copy ## of ## copies
USARPA
C FORT SHAFTER,
USA 071145Z May
2013
Message reference number 002

(U) TAB C (MEDICAL EVACUATION) TO APPENDIX 3 (ARMY HEALTH SYSTEM SUPPORT) ANNEX F (SUSTAINMENT) TO OPERATION ORDER (OPORD) 00-001, OPERATION LIBERATION

(U) References: OPORD 00-001

  a. (U) Refer to maps and overlays as listed in Appendix 2 (Operation Overlay) to Annex C (Operations).

  b. (U) Doctrinal reference include FM 4-02, ATP 4-02.2, and ATP 4-25.13.

(U) Time Zone Used Throughout the Order: Universal Time (ZULU)

1. **(U) Situation.** Refer to Annex F (Sustainment) to OPORD 00-001.

  a. (U) <u>Area of Interest.</u>   Refer to Annex B (Intelligence).

  b. (U) <u>Area or Operations.</u> Refer to Appendix 2 (Operation Overlay) to Annex C (Operations).

    (1) (U) <u>Terrain.</u>   Overall terrain is comprised of rolling hills with sparsely forested and build-up areas. The road network provides adequate means for ground ambulance movement.   Refer to Tab A (Terrain) to Appendix 1 (Intelligence Estimate) to Annex B (Intelligence) as required.

(page number)
(CLASSIFICATION)

Figure B-2. Example of a medical evacuation tab to an AHS appendix

---

(CLASSIFICATION)

(2) (U) <u>Weather</u>. Seasonal rains and associated low cloud ceiling may have adverse effects on air evacuation assets during night and day missions. Heavy rains may also make secondary roads impassable for extended periods. Refer to Tab B (Weather) to Appendix 1(Intelligence Estimate) to Annex B (Intelligence) as required.

c. (U) <u>Enemy Forces</u>. As addressed in OPORD 00-001, the ground-to-air threat is significant in all enemy sectors. The air-to-air threat is minimal, in all sectors. All planned ground evacuation routes will be targets for ambush by enemy forces and sympathetic host nationals.

d. (U) <u>Friendly Forces</u>. A six (6) aircraft (HH-60) Air Force combat search and rescue unit will be staged from the aerial port of debarkation. They will respond to downed aircrew evacuation recovery requests from the joint personnel recovery center.

e. (U) <u>Interagency, Intergovernmental, and Nongovernmental Organizations</u>. The Red Cross and Red Crystal are operating in the urban and rural areas providing ground ambulance support to and from displaced civilian camps (see Annex K [Civil Operations] for current locations) and civilian medical treatment facilities (MTFs).

f. U) <u>Civil Considerations.</u> Detainees and dislocated civilian will be evacuated as their medical evacuation precedence requires. Detainees require an armed escort that will be provided by requesting organization. Evacuation crews will not be responsible for providing escorts. Refer to Tab C (Civil Considerations) to Appendix 1(Intelligence Estimate) to Annex B (Intelligence) and Annex K (Civil Affairs Operations) as required.

g. (U) <u>Attachments and Detachments</u>. Refer to Annex A (Task Organization).

h. (U) <u>Assumptions</u>. None.

2. (U) **Mission**. On order provide tactical and operational ground and air intratheater medical evacuation or patient movement in support of Operation Liberation and continue coverage of all combined joint task force assets until relieved.

3. (U) **Execution**.

a. (U) <u>Scheme of Medical Evacuation Support</u>. This is a three phase operation.

(1) (U) <u>Phase 1</u>—Begins prior to initiation of the attack. All United States forces will be in the staging areas and in preattack positions. There will be no or very limited direct contact with enemy forces. C 2-1, C 2-2, and C 2-3 will not be required to provide evacuation support until D-3 days. Units in the staging area will be directed to call C 2-4 for evacuation at all times prior to D-3 days. All other evacuation units will concentrate on training and preparation for hostilities as directed by their commanders.

(2) (U) <u>Phase 2</u>—Begins at the start of the attack (H-Hour). C 2-1, C 2-2, C 2-3 will provide direct support to their assigned, area support to the Corps sustainment area and the backhaul mission.

(3) (U) <u>Phase 3</u>—Begins from the conclusion of the attack until stability task commence. Evacuation assets will begin transferring to an area support mission during this phase.

b. (U) <u>Task to Subordinate Units.</u>

(1) (U) <u>Phase 1.</u> C 2-1, C 2-2, C 2-3 will provide direct support to their division assets from D-3 on. C 2-1, C 2-2, C 2-3 will respond to any request for evacuation that is geographically close to their positions even if not in their division area. C 2-4 will provide support to the entire staging area (area of operations [AO] Buffalo). C 2-4 will perform the shore-to-ship and the ship-to-shore mission as directed by the 11th Medical Command (deployment support) (MEDCOM [DS]) theater patient movement center (TPMC). The 123rd MED CO (GA) will operate in the 1st Marine Expeditionary Force sector in a direct support role. 1st Medical Battalion (multi-functional) (MMB) will cover the 1st and 2nd Division footprint. The 234th MED CO (Ground ambulance) will provide support to Camps Jones, Smith, and Franks and perform patient transfers from camps to the aerial port of debarkation.

(page number)
(CLASSIFICATION)

Figure B-2. Example of a medical evacuation tab to an AHS appendix (continued)

(CLASSIFICATION)

(2) (U) <u>Phase 2.</u> C 2-1, C 2-2, C 2-3 will provide direct support to their division assets. C 2-1, C 2-2, C 2-3 will be responsible for moving patients from within the division area back to the staging AO Buffalo until forward line of own troops is beyond phase line Alpha. After the forward line of own troops is beyond phase line Alpha (at which time one way flight time for the H-60 back to AO Buffalo will exceed 2 hours), C 2-1 will designate one FSMP to stage at XYZ Airfield to perform the backhaul and the area support mission until relieved. United States Air Force aeromedical transports will be coordinated for by 11th MEDCOM (DS) flying from XYZ airfield. Beyond phase line Alpha C 2-2 will provide one FSMP to support the greater casualties expected in C 2-2 sector. Coordinate internally for link-up. C 2-4 will provide support to the entire staging area (AO Buffalo). C 2-4 will perform the ship-to-shore mission as directed by the 11th MEDCOM (DS) medical regulating office (MRO). C 2-4 will designate 4 aircraft to go forward as required to backhaul patients to AO Buffalo.

(3) (U) <u>Phase 3.</u> C 2-1, C 2-2, C 2-3 will provide direct support to their Divisions' assets. The 1st Marine Expeditionary Force will identify areas of high casualty density conducive to ground evacuation (urban areas) and divide the resources of the 123rd MED CO (GA) to cover these areas efficiently, if required. The 234th MED CO (GA) will conduct convoy support and provide coverage for AO Buffalo.

c. (U) <u>Coordinating Instructions.</u>

(1) (U) <u>Chemical, Biological, Radiological and Nuclear (CBRN) Conditions.</u> The preferred method of evacuating CBRN patients is by ground. Patients may be evacuated by air ambulance once decontamination is complete.

(2) (U) <u>Air and Ground Escort Requirements.</u> Ambulance crews will coordinate with supporting units for armed escorts. Air ambulances will follow theater guidance for dual ship and escort requirements. Waiver requirements for air ambulances will rest with the combat aviation brigade (CAB) commander and may be delegated down to the general support aviation battalion (GSAB) commander. Ground ambulances will follow theater convoy requirements and coordinate accordingly.

(3) (U) <u>Scheduled Movement of ROUTINE or CONVENIENCE Patients.</u> Movement of ROUTINE and CONVENIENCE patients from MTFs along secure routes of evacuation should be conducted using ground evacuation assets. All MMBs will establish patient movement schedules with supporting MTFs. When distances are beyond acceptable times and conditions for safe patient movement as determined by the sending physician, MMBs will coordinate with supporting CABs for air ambulance support.

(4) (U) <u>Medical Regulating.</u> The 11th MEDCOM (DS) will oversee medical regulating during all phases of the operation.

(5) (U) <u>Tentative Locations of MTFs by Phase.</u>

(a) (U) <u>Phase 1</u>: *2nd CSH AB 1234 5678, 1st CSH AB 2345 6789*

(b) (U) <u>Phase 2</u>: *2nd CSH AB 1234 5678, 1st CSH AB 2345 6789*

(c) (U) <u>Phase 3</u>: *2nd CSH AB 1234 5678, 1st CSH AB 2345 6789*

(6) (U) <u>Mass Casualty.</u> If a mass casualty event develops that overwhelms the resources available to a single MED CO (AA), contact the 11th MEDCOM (DS) for coordinating assistance.

(7) (U) <u>Casualty Collection.</u> Military treatment facilities can call in a medical evacuation mission request to any supporting medical evacuation asset for URGENT and PRIORITY missions, however MTFs that need to clear beds will coordinate mass patient movements through CJTF G-3 air for possible USAF support.

(page number)
(CLASSIFICATION)

Figure B-2. Example of a medical evacuation tab to an AHS appendix (continued)

Appendix B

> (CLASSIFICATION)
>
> (a) (U) 1st CSH will call C 2-4 for evacuation requests during Phase 1.
>
> (b) (U) 2nd CSH will call C 2-4 for evacuation requests during Phase 1.
>
> (c) (U) 1st CSH will call C 2-1 for evacuation requests during Phase 2/3.
>
> (d) (U) 2nd CSH will call C 2-2 for evacuation requests during Phase 2/3.
>
> (8) (U) Ground Evacuation. Seasonal monsoons are expected to washout many of the planned main supply routes and cause flooding.
>
> (9) (U) Air Evacuation. If air evacuation routes become so long as to seriously degrade the capabilities of movement by air ambulance, contact CJTF assistant chief of staff, operations Air for additional support and possible reallocation of evacuation assets. An air ambulance company is considered to be operating beyond its capability or inefficiently if—
>
> (a) (U) More than 50 percent of missions requiring a one-way trip time exceeding 2 hours.
>
> (b) (U) More than 50 percent of missions flown, are flown through another MED CO (AA) area of direct support coverage.
>
> (10) (U) Medical Considerations for Evacuation.
>
> (a) (U) United States Army medical evacuation assets will be used to evacuate all injured personnel on the battlefield without prejudice.
>
> (b) (U) Human immunodeficiency virus is at epidemic proportions in the Simula population. Appropriate precautions should be taken during treatment and when dealing with biological material.
>
> (c) (U) The medical evacuation of displaced civilians and detainees is estimated to make up as much as 75 percent of the total mission patient load.
>
> (11) (U) Commander's Critical Information Requirement (CCIR). Divisions G-3 will report the following CCIR to CJTF G-3 Air. Use the CJTF CCIR format.
>
> (a) (U) Any mass casualty.
>
> (b) (U) Any medical evacuation company with degraded capability due to evacuation distance as defined above in 3.d (9) (a)–(b).
>
> (c) (U) Any medical evacuation company with an operational readiness rate below 70 percent.
>
> 1. (U) Personnel: number of crews' less than available air or ground ambulance assets.
>
> 2. (U) Any loss of air or ground ambulances by enemy action or any violation of the Geneva Conventions.
>
> (d) (U) Casualty Estimates. Refer to Appendix 2 (Army Health System Support) to Annex F (Sustainment) for specific casualty estimates.
>
> (e) (U) United States Air Force Aeromedical Evacuation. Combined joint task force G-3 Air will coordinate for USAF intratheater aeromedical missions.
>
> **4. (U) Sustainment.**
>
> a. (U) Material and Services. Refer to Annex F (Sustainment).
>
> (U) Maintenance.
>
> (page number)
> **(CLASSIFICATION)**

Figure B-2. Example of a medical evacuation tab to an AHS appendix (continued)

Example of the Medical Evacuation Plan and Operations Order

(CLASSIFICATION)

(1) (U) <u>Air Ambulance</u>. Maintenance support will be per unit SOP except C 2-3 will receive aviation unit maintenance/aviation intermediate maintenance support from the 2-4 CAB. Priority of support will be according to Annex F (Sustainment).

(2) (U) <u>Ground Ambulance</u>. Priority of maintenance effort will be in accordance with (IAW) Annex F (Sustainment).

   b. (U) <u>Supply</u>. Refer to Annex F (Sustainment).

(U) <u>Class VIII Medical Supply</u>. Refer to Annex F (Sustainment) as required.

5. (U) Command and Signal.

   a. (U) <u>Command</u>.

(1) (U) <u>Location of Commander</u>. See Appendix 2 (Operational Overlay) of Annex C (Operations) for unit command post locations.

(2) (U) <u>Succession of Command</u>. As per unit SOP.

(3) (U) <u>Liaison Requirements</u>. MED CO (AA) will provide liaisons to supported brigade combat teams (BCTs) as required.

   b. (U) <u>Control</u>.

(1) (U) <u>Command Posts</u>. See Appendix 2 (Operational Overlay) of Annex C (Operations) for unit command post locations.

(2) (U) <u>Reports</u>. Medical evacuation units are required to submit a report of evacuation missions not later than (NLT) 1400 daily to the 11th MEDCOM (DS). The report will show number of patients by evacuation precedence and mode of evacuation. Medical evacuation units assigned to a division will submit their daily report of evacuation missions by 1200 to the division surgeon section. Divisions will submit the reports to the 11th MEDCOM (DS) the theater medical command not later than 1400.

   c. (U) <u>Signal</u>.

(1) (U) Combined joint task force surgeon tactical chat call sign, CJTF_SURG_EVAC

(2) (U) 11th MEDCOM (DS) tactical chat call sign, 11MED_G-3_EVAC

(3) (U) 1st MEDBDE (SPT) tactical chat call sign, 1MEDBDE_OPS_EVAC

(4) (U) 1st Division surgeon, tactical chat call sign, 1DIV_G4_SURG

(5) (U) 2nd Division Surgeon (G-3), tactical chat call sign, 2DIV_G4_SURG

(6) (U) 1 CAB medical evacuation cell, tactical chat call sign, 1CAB_OPS_EVAC

(7) (U) 2 CAB medical evacuation cell, tactical chat call sign, 2CAB_OPS_EVAC

(8) (U) 1st Marine Expeditionary Force, medical evacuation cell, tactical chat call sign, 1MEF_OPS_EVAC

ACKNOWLEDGE: **PATRICK**
                                                               LTG

FOR THE COMMANDER
OFFICIAL:

GORDON
COL

(page number)
(CLASSIFICATION)

Figure B-2. Example of a medical evacuation tab to an AHS appendix (continued)

This page intentionally left blank.

Appendix C
Example of Medical Evacuation Request

This appendix provides an example of the 9-line medical evacuation request in Table C-1 below. The same format is used for both air and ground medical evacuation requests.

Table C-1. 9-Line Medical Evacuation Request

Line	Item	Explanation	Where/how obtained	Who normally provides	Reason
1	Location of pickup site.	Grid coordinates of the pickup site should be sent by secure communication. To prevent confusion the grid zone letters are included in the message.	From map or navigational device determine the military grid reference system six-digit grid coordinates of the pickup site.	Unit leader(s).	Required so evacuation vehicle knows where to pick up the patient/casualty. Also, so that the unit coordinating the evacuation mission can plan the route for the evacuation vehicle (if the evacuation vehicle must pick up from more than one location).
2	Radio frequency, call sign and suffix.	Frequency of the radio at the pickup site, not a relay frequency. The call sign (and suffix if used) of person to be contacted at the pickup site may be transmitted in the clear.	From automated net control device or other approved means.	Radio transmission operator.	Required so that evacuation vehicle can contact requesting unit while en route (obtain additional information or changes in situation or directions).
3	Number of patients by precedence.	A—URGENT B—URGENT-SURG C—PRIORITY D—ROUTINE E—CONVENIENCE If two or more categories must be reported in the same request, insert the word "BREAK" between each category.	From evaluation of patients.	Medic or senior person present.	Required by unit controlling vehicles to assist in prioritizing missions.
4	Special equipment required.	A—None B—Hoist C—Extraction equipment D—Ventilator	From evaluation of patient/ situation.	Medic or senior person present.	Required so that the equipment can be placed on board the evacuation vehicle prior to the start of the mission.

Table C-1. 9-Line Medical Evacuation Request (continued)

Line	Item	Explanation	Where/how obtained	Who normally provides	Reason
5	Number of patients by type.	Report only applicable information, if requesting medical evacuation for both types, insert the word "BREAK" between the litter entry and ambulatory entry. L+# of patients–Litter A+# of patients–Ambulatory (sitting)	From evaluation of patients.	Medic or senior person present.	Required so that the appropriate number of evacuation vehicles may be dispatched to the pickup site. They should be configured to carry the patients requiring evacuation.
6	Security of pickup site (wartime).	N—No enemy troops in area. P—Possibly enemy troops in area (approach with caution). E—Enemy troops in area (approach with caution). X—Enemy troops in area (armed escort required).	From evaluation of situation.	Unit leader.	Required to assist the evacuation crew in assessing the situation and determining if assistance is required. More definitive guidance can be furnished to the evacuation vehicle while it is en route (specific location of enemy to assist an aircraft in planning its approach).
6	Number and type of wound, injury or illness (peacetime).	Specific information regarding patient wounds by type (gunshot or shrapnel). Report serious bleeding, along with patient's blood type, if known.	From evaluation of patients.	Medic or senior person present.	Required to assist evacuation personnel in determining treatment and special equipment needed.
7	Method of marking pickup site.	A—Panels B—Pyrotechnic signal C—Smoke signal D—None E—Other	Based on situation and availability of materials.	Medic or senior person present.	Required to assist the evacuation crew in identifying the specific location of the pickup. Note that the color of the panel or smoke should not be transmitted until the evacuation vehicle contacts the unit (just prior to its arrival). For security, the crew should identify the color and the unit verifies it.
8	Patient nationality and status.	The number of patients in each category need not be transmitted. A—U.S. military B—U.S. citizen C—Non-U.S. military D—Non-U.S. citizen E—enemy prisoner of war	From evacuation platform.	Medic or senior person present.	Required to assist in planning for destination facilities and need for guards. Unit requesting support should ensure that there is an English-speaking representative at the pickup site.

Table C-1. 9-Line Medical Evacuation Request (continued)

Line	Item	Explanation	Where/how obtained	Who normally provides	Reason
9	Chemical, Biological, Radiological, and Nuclear contamination (wartime).	Include this line only when applicable C—Chemical B—Biological R—Radiological N—Nuclear	From situation.	Medic or senior person present.	Required to assist in planning for the mission. (Determine which evacuation vehicle will accomplish the mission and when it will be accomplished.)
9	Terrain description (peacetime).	Includes details of terrain features in and around proposed landing site. If possible, describe relationship of site to prominent terrain feature (lake, mountain, or tower).	From area survey.	Personnel present.	Required to allow evacuation personnel to assess route/avenue of approach into area. Of particular importance if hoist operation is required.

This page intentionally left blank.

Appendix D
Example of Medical Evacuation Activities During Operations

The following figures provide examples of how medical evacuation units may be allocated to evacuate patients during operations to shape, prevent, large-scale combat operations, and consolidation of gains.

THE PROVISION OF MEDICAL EVACUATION SUPPORT DURING OPERATIONS

D-1. Medical evacuation support to operations may be conducted differently based on mission variables and other factors. While the MEDEVAC tasks and attributes remain constant, these figures provide "a way" but not "the only way" MEDEVAC support is provided.

MEDICAL EVACUATION DURING OPERATIONS TO SHAPE

D-2. Operations to shape consist of various long-term military engagements, security cooperation, and deterrence missions, tasks, and actions intended to assure friends, build partner capacity and capability, and promote regional stability (FM 3-0).

D-3. Medical evacuation units will continue to train on individual, collective, mission essential task list (commonly known as METL) tasks, conduct clinical training and rotations to sustain medical skills and certifications, and pursue professional development courses.

D-4. Staff exercises may be held at tactical through operational levels and units may prepare time-phased and deployment data, update equipment sets, and prepare containers and vehicles for deployment.

D-5. Key shaping activities may include support to military exercises. Medical evacuation units can support security cooperation and security force assistance goals through activities such as medical evacuation training to build partner capacity.

D-6. Figure D-1 on page D-2 depicts MEDEVAC units and the movement of patients from POI to Role 4 in support of operations to shape.

Appendix D

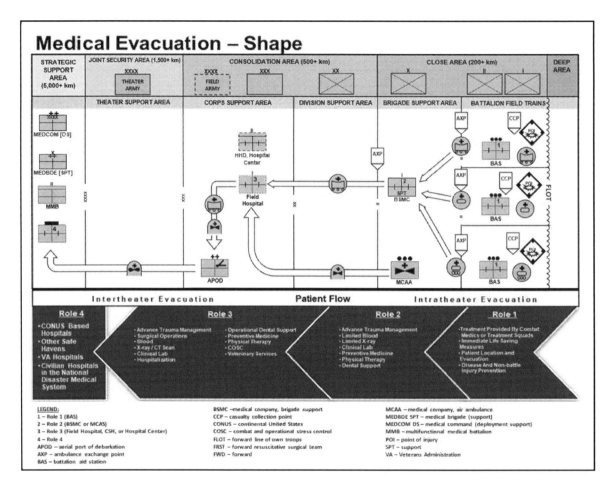

Figure D-1. Medical evacuation support during operations to shape

MEDICAL EVACUATION DURING OPERATIONS TO PREVENT

D-7. The purpose of operations to prevent is to deter adversary actions contrary to U.S. interests (FM 3-0).

D-8. At the strategic level, the theater army plans and coordinates Army capabilities to set the theater. Planners assess available intelligence to inform medical planning and develop contingency or operation plans.

D-9. Medical planners should begin planning early, develop an understanding of the available Army, Joint, and Coalition forces, identify requirements, and how to best provide support with the available medical evacuation assets which may include units and capabilities from other Services.

D-10. Units and personnel who are part of a readiness force or who have been designated for specific operation will have completed clinical rotations and training events and be available for immediate recall and deployment. Medical evacuation units may provide area MEDEVAC support during reception, staging, onward movement and integration, at ports of embarkation, debarkation, and along movement routes. A rapid deployment as part of operations to prevent will likely involve only tactical MEDEVAC capabilities such as those during airborne operations.

D-11. Figure D-2 depicts MEDEVAC units and the movement of patients from POI to Role 4 in support of operations to prevent.

Example of the Medical Evacuation Activities During Operations

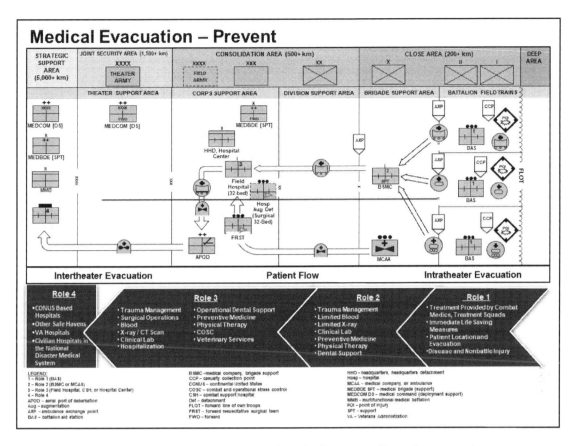

Figure D-2. Medical evacuation during operations to prevent

Appendix D

MEDICAL EVACUATION DURING LARGE-SCALE COMBAT OPERATIONS

D-12. Medical evacuation activities during large-scale combat operations are discussed in Chapter 2 of this publication. The complexity and lethality of the environment will require MEDEVAC units to operate across multiple domains in a synchronized effort with the MTFs to clear the battlefield thereby sustaining the initiative of the maneuver commander. Figure D-3 depicts MEDEVAC units and the movement of patients from POI to Role 4 in support of large-scale combat operations.

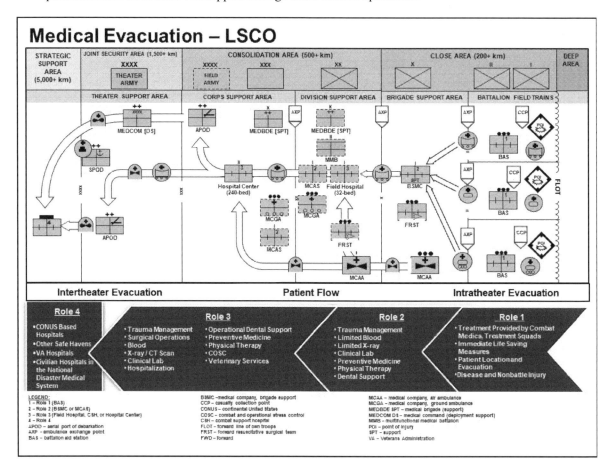

Figure D-3. Medical evacuation during large-scale combat operations

MEDICAL EVACUATION DURING OPERATIONS TO CONSOLIDATE GAINS

D-13. Consolidate gains activities occur within sections of an area of operations where large-scale combat operations are no longer occurring and consists of security and stability tasks. Combat against remnant or bypassed enemy forces presents a threat to MEDEVAC units and teams utilizing evacuation routes or AXPs due to their vulnerability and lack of offensive capability. Armed escort vehicles serve as a deterrent to attack and provide increased defensive firepower.

D-14. Medical evacuation support during operations to consolidate gains may require a MEDEVAC company to provide direct support to maneuver forces in one area while supporting stability tasks in another.

D-15. Figure D-4 depicts MEDEVAC units and the movement of patients from POI to Role 4 in support of operations to consolidate gains.

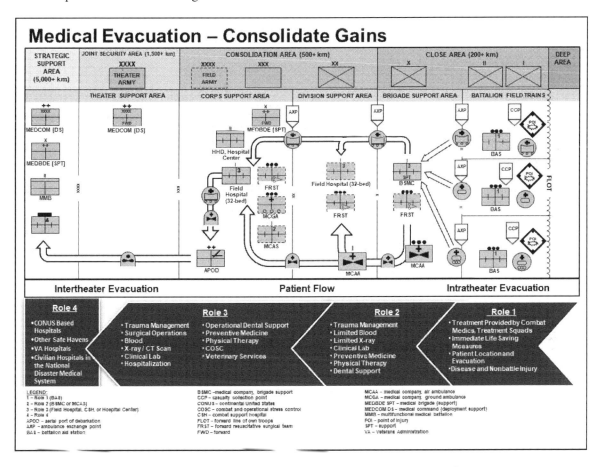

Figure D-4. Medical evacuation during consolidation of gains

This page intentionally left blank.

Glossary

This glossary lists acronyms and terms with Army or joint definitions. Where Army and joint definitions differ, (Army) precedes the definition. Terms for which ATP 4-02.2 is the proponent are marked with an asterisk (*). The proponent publication for other terms is listed in parentheses after the definition.

SECTION I — ACRONYMS AND ABBREVIATIONS

ADP	Army doctrine publication
ADRP	Army doctrine reference publication
AECT	aeromedical evacuation control team
AELT	aeromedical evacuation liaison team
AFTTP	Air Force tactics, techniques, and procedures
AHS	Army Health System
AO	area of operations
AR	Army regulation
ARSOF	Army special operations forces
ASMP	Area support medical evacuation platoon
ATP	Army techniques publication
ATTP	Army tactics, techniques, and procedures
AXP	ambulance exchange point
BAS	battalion aid station
BCT	brigade combat team
BSB	brigade support battalion
CAB	combat aviation brigade
CASEVAC	casualty evacuation
CBRN	chemical, biological, radiological, and nuclear
CCDR	combatant commander
CCP	casualty collection point
CJTF	combined joint task force
CONUS	continental United States
CSH	combat support hospital
DD	Department of Defense (Forms only)
DOD	Department of Defense
DODD	Department of Defense directive
DODI	Department of Defense instruction
DSCA	defense support of civil authorities
EAB	echelons above brigade
FHP	force health protection

FM	field manual
FRST	forward resuscitative surgical team
FSMP	Forward support medical evacuation platoon
FST	forward surgical team
G-3	assistant chief of staff, operations
GA	ground ambulance
GPMRC	global patient movement requirements center
GSAB	general support aviation battalion
GWS	Geneva Convention for the Amelioration of the Condition of the Wounded and Sick in Armed Forces in the Field
HEEDS	Helicopter Emergency Egress Device System
HMMWV	high mobility multipurpose wheeled vehicle
HSS	health service support
JP	joint publication
JPMRC	joint patient movement requirements center
LZ	landing zone
MCAS	medical company (area support)
MCRP	Marine Corps reference publication
MCTP	Marine Corps techniques publication
MCWP	Marine Corps warfighting publication
MEDBDE (SPT)	medical brigade (support)
MEDCOM (DS)	medical command (deployment support)
MEDEVAC	medical evacuation
METT-TC	mission, enemy, terrain and weather, troops and support available, time available, and civil considerations
MIST	mechanism of injury, injury type, signs, treatment
MMB	medical battalion (multifunctional)
MRO	medical regulating officer
MTF	medical treatment facility
NATO	North Atlantic Treaty Organization
NTTP	Navy tactics, techniques, and procedures
OPCON	operational control
OPLAN	operation plan
OPORD	operation order
PAD	patient administration division
PMI	patient movement items
POI	point of injury
S-3	Operations staff officer
SOF	Special operations forces
SOP	standard operating procedures
STANAG	standardization agreement
TAES	Theater Aeromedical Evacuation System

TCCC	tactical combat casualty care
TPMC	theater patient movement center
TPMRC	Theater patient movement requirements center
TRAC2ES	Transportation Command regulating and Command and Control Evacuation System.
TPMRC	theater patient movement requirements center
U.S.	United States
USAF	United States Air Force
USMC	United States Marine Corps
USTRANSCOM	United States Transportation Command
XO	executive officer

SECTION II — TERMS

aeromedical evacuation

> The movement of patients under medical supervision to and between medical treatment facilities by air transportation. Also called AE. (JP 4-02)

*ambulance control point

> A manned traffic regulating, often stationed at a crossroad or road junction, where ambulances are directed to one of two or more directions to reach loading points and medical treatment facilities.

*ambulance exchange point

> A location where a patient is transferred from one ambulance to another en route to a medical treatment facility. Also called AXP.

*ambulance loading point

> This is the point in the shuttle system where one or more ambulances are stationed ready to receive patients for evacuation.

*ambulance relay point

> A point in the shuttle system where one or more empty ambulances are stationed to advance to a loading point or to the next relay post to replace departed ambulances.

*ambulance shuttle system

> A system consisting of one or more ambulance loading points, relay points, and when necessary, ambulance control points, all echeloned forward from the principal group of ambulances, the company location, or basic relay points as tactically required.

area of operations

> An operational area defined by a commander for land and maritime forces that should be large enough to accomplish their missions and protect their forces. Also called AO. (JP 3-0)

area support

> (Army) Method of logistics, medical support, and personnel services in which support relationships are determined by the location of the units requiring support. Sustainment units provide support to units located in or passing through their assigned areas. (ATP 4-90)

casualty

> Any person who is lost to the organization by having been declared dead, duty status — whereabouts unknown, missing, ill, or injured. (JP 4-02)

*casualty collection point

> A location that may or may not be staffed, where casualties are assembled for evacuation to a medical treatment facility.

Glossary

casualty evacuation
> The movement of casualties aboard nonmedical vehicles or aircraft without en route medical care. Also called CASEVAC. (ATP 4-25.13)

direct support
> (Army) A support relationship requiring a force to support another specific force and authorizing it to answer directly to the supported force's request for assistance. (FM 3-0)

*en route care
> The care required to maintain the phased treatment initiated prior to evacuation and the sustainment of the patient's medical condition during evacuation.

*lines of patient drift
> Natural routes along which wounded Soldiers may be expected to go back for medical care from a combat postion.

mass casualty
> Any number of human casualties produced across a period of time that exceeds available medical support capabilities. (JP 4-02)

*medical evacuation
> The timely and effective movement of the wounded, injured, or ill to and between medical treatment facilities on dedicated and properly marked medical platforms with en route care provided by medical personnel. Also called MEDEVAC.

medical regulating
> The actions and coordination necessary to arrange for the movement of patients through the roles of care and to match patients with a medical treatment facility that has the necessary health service support capabilities and available bed space. (JP 4-02)

*nontransportable patient
> A patient whose medical condition is such that he could not survive further evacuation to the rear without surgical intervention to stabilize his medical condition.

patient
> A sick, injured or wounded Soldier who receives medical care or treatment from medically trained personnel. (FM 4-02)

*patient movement
> The act of moving a sick, injured, wounded, or other person to obtain medical and/or dental treatment.

*theater evacuation policy
> A command decision indicating the length in days of the maximum period of noneffectiveness that patients may be held within the command for treatment, and the medical determination of patients that cannot return to duty status within the period prescribed requiring evacuation by the first available means, provided the travel involved will not aggravate their disabilities or medical condition.

References

All URLs were accessed on 19 April 2019.

REQUIRED PUBLICATIONS

These documents must be available to the intended users of this publication.

These publications are available online at: https://armypubs.army.mil/.

ADP 1-02, *Terms and Military Symbols*, 14 August 2018.

FM 4-02, *Army Health System*, 26 August 2013.

These publications are available online at: http://www.jcs.mil/Doctrine/.

DOD Dictionary of Military and Associated Terms, May 2019.

RELATED PUBLICATIONS

These documents contain relevant supplemental information.

AMERICAN, BRITISH, CANADIAN, AUSTRALIAN AND NEW ZEALAND (ARMIES) STANDARDS

These documents are available online at https://www.apan.org/. (Must create an account to access ABCA Standards.)

Standard 436, *Minimum Labelling Requirements for Medical Materiel*, Edition 1, 31 January 1979.

Standard 2079, *ABCA Patient Medical Evacuation Request*, 13 August 2010.

Standard 2080, *Coalition Casualty Regulating Tool (CRRT)*, 2 December 2009.

ARMY PUBLICATIONS

Most Army doctrinal publication are available online at https://armypubs.army.mil/.

ADP 3-07, *Stability*, 31 August 2012.

ADP 3-28, *Defense Support of Civil Authorities*, 11 February 2019.

ADP 3-90, *Offense and Defense*, 13 August 2018.

ADP 4-0, *Sustainment*, 31 July 2012.

ADP 5-0, *The Operations Process*, 17 May 2012.

ADP 6-0, *Mission Command*, 17 May 2012.

ADRP 3-05, *Special Operations*, 29 January 2018.

AR 40-3, *Medical, Dental, and Veterinary Care*, 23 April 2013.

AR 95-1, *Flight Regulations*, 22 March 2018.

AR 380-5, *Department of the Army Information Security Program*, 29 September 2000.

AR 750-1, *Army Material Maintenance Policy*, 3 August 2017.

ATP 3-04.1, *Aviation Tactical Employment*, 13 April 2016.

ATP 3-11.50, *Battlefield Obscuration*, 15 May 2014.

ATP 3-90.97, *Mountain Warfare and Cold Weather Operations*, 29 April 2016.

ATP 4-02.1, *Army Medical Logistics*, 29 October 2015.

ATP 4-02.3, *Army Health System Support to Maneuver Forces*, 9 June 2014.

ATP 4-02.5, *Casualty Care*, 10 May 2013.

References

ATP 4-02.42, *Army Health System Support to Stability and Defense Support of Civil Authorities Tasks*, 9 June 2014.

ATP 4-02.43, *Army Health System Support to Army Special Operations Forces*, 17 December 2015.

ATP 4-02.55, *Army Health System Support Planning*, 16 September 2015.

ATP 4-25.13, *Casualty Evacuation*, 15 February 2013.

ATP 4-90, *Brigade Support Battalion*, 2 April 2014.

FM 1-564, *Shipboard Operations*, 29 June 1997.

FM 3-0, *Operations*, 6 October 2017.

FM 3-04, *Army Aviation*, 29 July 2015.

FM 3-21.38, *Pathfinder Operations*, 25 April 2006.

FM 3-50, *Army Personnel Recovery*, 2 September 2014.

FM 3-99, *Airborne and Air Assault Operations*, 6 March 2015.

FM 6-0, *Commander and Staff Organization and Operations*, 5 May 2014.

FM 27-10, *The Law of Land Warfare*, 18 July 1956.

FM 90-5, *Jungle Operations*, 16 August 1982.

DEPARTMENT OF DEFENSE PUBLICATIONS

Department of Defense publications are available online at http://www.esd.whs.mil/DD/.

Department of Defense Law of War Manual, 12 June 2015.

https://dod.defense.gov/Portals/1/Documents/law_war_manual15.pdf

DODD 3025.18, *Defense Support of Civil Authorities (DSCA)*, 29 December 2010.

DODD 5100.01, *Functions of the Department of Defense and Its Major Components*, 21 December 2010.

DODI 6000.11, *Patient Movement (PM)*, 22 June 2018.

GENEVA CONVENTIONS

These documents are available online at https://www.loc.gov/rr/frd/Military_Law/pdf/GC_1949-I.pdf.

Convention (I) for the Amelioration of the Condition of the Wounded and Sick in Armed Forces in the Field, 12 August 1949.

JOINT PUBLICATIONS

These publications are available online at http://www.jcs.mil/doctrine/.

JP 3-0, *Joint Operations*, 17 January 2017.

JP 4-02, *Joint Health Service*, 11 December 2017.

JP 5-0, *Joint Planning*, 16 June 2017.

MULTISERVICE PUBLICATIONS

Most multi-Service doctrinal publications are available online at https://armypubs.army.mil/.

ATP 3-06/MCTP 12-10B, *Urban Operations*, 7 December 2017.

ATP 3-06.1/MCP 3-35.3A/NTTP 3-01.04/AFTTP 3-2.29, *Aviation Urban Operations (Multi-Service Tactics, Techniques, and Procedures for Aviation Urban Operations)*, 27 April 2016.

ATP 3-11.32/MCWP 10-10E.8/NTTP 3-11.37/AFTTP 3-2.46, *Multi-Service Tactics, Techniques, and Procedures for Chemical, Biological, Radiological, and Nuclear Passive Defense*, 13 May 2016.

ATP 4-02.7/MCRP 4-11.1F/NTTP 4-02.7/AFTTP 3-42.3, *Multi-Service Tactics, Techniques, and Procedures for Health Service Support in a Chemical, Biological, Radiological, and Nuclear Environment*, 15 March 2016.

NORTH ATLANTIC TREATY ORGANIZATION STANDARDIZATION AGREEMENTS

These documents are available online at: https://nso.nato.int/. (Must request access to protected site.)

STANAG 2040, *Stretchers, Bearing Brackets and Attachment Supports,* Edition 7, 18 November 2013 (AMedP-2.1, Edition A, Version 1, 18 November 2013).

STANAG 2087, *Forward Aeromedical Evacuation,* Edition 7, 23 April 2018 (AAMedP-1.5, Edition A, Version, Edition 6, 23 April 2018).

STANAG 2128, *Medical and Dental Supply Procedures,* Edition 6, 21 January 2014 (AMedP-1.12, Edition A, Version 1, 21 January 2014).

STANAG 2132, *Documentation Relative to Initial Medical Treatment and Evacuation,* Edition 3, 11 June 2013 (AMedP-8.1, Edition A, Version 2, 6 September 2016).

STANAG 2454, *Road Movements and Movement Control,* Edition 3, 27 January 2005 (AMovP-1[A], December 2004).

STANAG 2931, *Orders for the Camouflage of Protective Medical Emblems on Land in Tactical Operations,* Edition 4, 19 January 2018 *(*ATP-79*,* Edition B, 19 January 2018).

STANAG 3204, *Aeromedical Evacuation,* Edition 8, 17 November 2014 (AAMedP-1.1, Edition A, Version 1, 17 November 2014).

OTHER DOCUMENTS AND PUBLICATIONS

National Security Strategy. Office of the President of the United States, 2017.
https://www.whitehouse.gov/wp-content/uploads/2017/12/NSS-Final-12-18-2017-0905-2.pdf.

National Strategy for Homeland Security, Homeland Security Council, October 2007.
https://www.dhs.gov/xlibrary/assets/nat_strat_homelandsecurity_2007.pdf.

UNITED STATES AIR FORCE PUBLICATIONS

This publication is available online at http://www.e-publishing.af.mil.

AFTTP 3-42.57, *En Route Patient Staging System,* 10 August 2016.

UNITED STATES LAW

This document is available online at http://uscode.house.gov.

Title 32, United States Code, National Guard.

WEB SITES

Aviation and Missile Command. https://asmprd.redstone.army.mil. (Safety of Flight [SOF], Technical, Rescue Hoist Operational Restrictions. Select Consolidated Listing→Aviation→H-47→SOF→Search→H-47-16-SOF-04.)

PRESCRIBED FORMS

This section contains no entries.

REFERENCED FORMS

Unless otherwise indicated, DA forms are available on the Army Publishing Directorate Web site: https://armypubs.army.mil.

DA Form 2028, *Recommended Changes to Publications and Blank Forms.*

Department of Defense (DD) Forms are available on the Office of the Secretary of Defense Web site: http://www.esd.whs.mil/DD/.

DD Form 1380, *Tactical Combat Casualty Care (TCCC) Card.*
Instructions for completing DD Form 1380 can be found at:
https://www.esd.whs.mil/Portals/54/Documents/DD/forms/dd/dd1380inst.pdf.

RECOMMENDED READINGS

These readings contain relevant supplemental information.

Most Army doctrinal publication are available online at https://armypubs.army.mil/.

ADP 3-05, *Special Operations*, 29 January 2018.

AFTTP 3-42.5, *Tactical Doctrine, Aeromedical Evacuation (AE)*, 1 November 2003.

AR 71-32, *Force Development and Documentation Consolidated Policies*, 20 March 2019.

ATTP 3-06.11, *Combined Arms Operations in Urban Terrain*, 10 June 2011.

JP 3-17, *Air Mobility Operations*, 5 February 2019.

JP 3-18, *Joint Forcible Entry Operations*, 11 May 2017.

JP 3-27, *Homeland Defense*, 10 April 2018.

JP 3-28, *Defense Support of Civil Authorities*, 29 October 2018.

Index

References are to paragraph numbers unless otherwise stated.

A

air assault, 2-69, 2-84–2-85, 2-88–2-94

airborne operations, 2-69, 2-86–2-87, 6-30, D-10

ambulance, 1-21–1-28, 2-23, 2-30, 2-33–2-36, 2-39, 2-40–2-42, 2-47, 2-53, 2-82–2-83 2-101, 2-105, Figure B-1, Figure B-2
   control point, 1-22–1-28, 5-26
   exchange point, 1-21–1-28 Figure 2-2, Figure B-1
   shuttle system, 1-25–1-27, 2-113, 2-157, Figure B-1

Army Health System, Introduction, Chapter 1, 2-42, 3-24, 4-6, 4-16, 5-3, 5-13, 5-15, 5-25, Table 5-1, Figure B-1–Figure B-2

B

battalion aid station, 1-19, 1-23, 1-27, 2-3, 2-15, 2-37, 2-39–42, Figure 2-2

C

camouflage, 2-183, A-4–A-5

casualty
   collection point, Introduction 1-19–1-20, 1-27, 2-15, 2-120, 5-26
   evacuation, Preface, Introduction, 1-29, 1-33, 1-37, 2-32

chemical, biological radiological, and nuclear, 2-80, 2-99, 2-175–2-177, Figure B-1–Figure B-2, Table C-1

continuum of care, Preface, 1-3, 1-9, 1-15–1-16, 1-18, 5-17, 6-6

D

detainee, 2-122, 2-151–2-153, A-16–A-17, Table 5-1, Figure B-1–Figure B-2

E

en route medical care, Preface, Introduction, 1-3, 1-5, 1-9, Table 1-1, 1-12–1-13, 1-16, 1-33–1-34, 2-39, 2-106, 3-28

enemy prisoners of war, Table 5-1

environmental
   CBRN operations, 2-175–2-176
   cold region operations, 2-172
   desert operations, 2-168
   jungle operations, 2-161–2-162
   mountain operations, 2-155

evacuation
   policy, Introduction, 1-7, 2-13, Chapter 4
   precedence, 2-1–2-3, Table 2-1,
   request, Preface, 2-7, 2-10, 2-14, Figure 2-1–Figure 2-2–Figure 2-3–Figure 2-4,Table C-1

G

Geneva Conventions, Introduction, 1-13, 2-1, 2-145, 2-182, Appendix A, Figure B-2

H

Hague Conventions, Appendix A

hoist rescue operations, 2-185, 2-187

homeland defense, 2-131, 2-137–2-138

J

joint
   operations, 1-16, 2-149, 5-1, 5-17, Figure B-1
   planning, 5-1–5-2,

M

mass casualty, 2-32, 2-80, Figure B-2

medical company
   air ambulance, 3-29–3-35, 3-39–3-44, Figure B-2

area support, 2-13, 2-16, Table 5-1, Figure B-2

brigade support battalion, 1-23, 2-15, 3-11

ground ambulance, 3-18–3-22, 5-5, Figure B-2

medical regulating, Preface, 1-2–1-8, Chapter 6, Figure B-1, Figure B-2

O

obscurants (use/employment of), 2-179
   Geneva Conventions, 2-180–2-186
   ground evacuation operations, 2-188
   hoist operations (signal smoke), 2-184–2-186
   landing zone marking (signal smoke), 2-183, Table C-1
   overwater operations (signal smoke), 2-187

P

patient movement items, 1-28, 2-55, Figure B-1

S

special operation forces, 2-106–2-108, 6-39–6-42

T

tactical combat casualty care 2-41, 2-99, 2-108, 3-7, 3-17

U

urban operations, 2-113–2-114, 2-116, 2-122–2-123, Figure B-1

Made in the USA
San Bernardino, CA
29 August 2019